DISTURBING THE UNIVERSE

THIS BOOK IS PUBLISHED AS PART
OF AN ALFRED P. SLOAN FOUNDATION PROGRAM

FREEMAN DYSON

DISTURBING THE UNIVERSE

BASIC
BOOKS

A Member of the Perseus Books Group

Portions of this work originally appeared in *The New Yorker*

Published by Basic Books,
A Member of the Perseus Books Group.

Basic Books, 10 E. 53rd Street, New York, NY 10022-5299

Library of Congress in Publication Data

Dyson, Freeman J.
    Disturbing the universe.
    Includes bibliographical references and index.
    1. Dyson, Freeman J.   2. Physicists—United States—
Biography.   3. Science.   I. Title.
OC16.D95A33   1979   530'.092'4 [B]   78-20665

ISBN   0–465–01677–4

Designed by Sidney Feinberg
10   9   8   7   6   5   4   3   2   1

To the undergraduates of the Universities of Idaho, California–Riverside, Emory, Furman, San Diego State and Florida State, Reed College and Smith College, with whom I talked as Phi Beta Kappa Visiting Scholar in 1975–76. They asked the questions which this book tries to answer.

Contents

III. POINTS BEYOND

Preface to the Series

The Alfred P. Sloan Foundation has for many years included in its areas of interest the encouragement of a public understanding of science. It is an area in which it is most difficult to spend money effectively. Science in this century has become a complex endeavor. Scientific statements are embedded in a context that may look back over as many as four centuries of cunning experiment and elaborate theory; they are as likely as not to be expressible only in the language of advanced mathematics. The goal of a general public understanding of science, which may have been reasonable a hundred years ago, is perhaps by now chimerical.

Yet an understanding of the scientific enterprise, as distinct from the data and concepts and theories of science itself, is certainly within the grasp of us all. It is, after all, an enterprise conducted by men and women who might be our neighbors, going to and from their workplaces day by day, stimulated by hopes and purposes that are common to all of us, rewarded as most of us are by occasional successes and distressed by occasional setbacks. It is an enterprise with its own rules and customs, but an understanding of that enterprise is accessible to any of us, for it is quintessentially human. And an understanding of the enterprise inevitably brings with it some insight into the nature of its products.

Accordingly, the Sloan Foundation has set out to encourage a representative selection of accomplished and articulate scientists to set down their own accounts of their lives in science. The form those accounts will take has been left in each instance to the author: one may choose an autobiographical approach, another may produce a coherent series of essays, a third may tell the tale of a scientific community of which he was a member. Each author is a man or woman of outstanding accomplishment in his or her field. The word "science" is not construed narrowly: it includes such disciplines as economics and anthropology as much as it includes physics and chemistry and biology.

The Foundation's role has been to organize the program and to provide the financial support necessary to bring manuscripts to completion. The Foundation wishes to express its appreciation of the great and continuing contribution made to the program by its Advisory Committee chaired by Dr. Robert Sinsheimer, Chancellor of the University of California–Santa Cruz, and comprising Dr. Howard H. Hiatt, Dean of the Harvard School of Public Health; Dr. Mark Kac, Professor of Mathematics at Rockefeller University; Dr. Daniel McFadden, Professor of Economics at the Massachusetts Institute of Technology; Robert K. Merton, University Professor, Columbia University; Dr. George A. Miller, Professor of Experimental Psychology at Rockefeller University; Professor Philip Morrison of the Massachusetts Institute of Technology; Dr. Frederick E. Terman, Provost Emeritus, Stanford University; for the Foundation, Arthur L. Singer, Jr., and Stephen White; for Harper & Row, Winthrop Knowlton and Simon Michael Bessie.

Author's Preface

The physicist Leo Szilard once announced to his friend Hans Bethe that he was thinking of keeping a diary: "I don't intend to publish it; I am merely going to record the facts for the information of God." "Don't you think God knows the facts?" Bethe asked. "Yes," said Szilard. "He knows the facts, but He does not know *this version of the facts.*"

I have collected in this book memories extending over fifty years. I am well aware that memory is unreliable. It not only selects and rearranges the facts of our lives, but also embroiders and invents. I have checked my version of the facts wherever possible against other people's memories and against written documents. For thirty years I wrote home regularly to my parents, and they kept most of my letters. These letters are the source of many details which memory alone could not have preserved.

I am grateful to the Alfred P. Sloan Foundation for funding the Science Book Program, under whose auspices this book appears. I thank Sloan Foundation Vice-President Stephen White and the members of his advisory committee for inviting me to write the book and for their editorial guidance. I am indebted for help and criticism to many friends, including Eileen Bernal, Jeremy Bernstein, Simon Michael Bessie, Hal Feiveson, Muguette Josefsen, Matthew Meselson, Mike O'Loughlin, Peter Partner, Leonard Rodberg, Barbara Scott, Martin Sherwin, Massoud Simnad, Daniel and Maxine Singer, Ted Taylor, Janet Whitcut, and my family. Above all I am grateful to my secretary, Paula Bozzay, for typing and retyping the manuscript.

Parts of chapters 10, 11, 12, 13 and 18 have appeared in print before. Detailed references will be found in the bibliographical notes at the end of the book.

I. ENGLAND

Oh England! Oh my lovely casual country!
Serenity of meadowland in April—
Carelessly littered with fritillaries,
Ladysmock, kingcups, cowslips, and wild apple!

FRANK THOMPSON, 1943

And there's a dreadful law here—it was made
by mistake, but there it is—that if any one asks
for machinery they have to have it and keep
on using it.

E. NESBIT, 1910

1

The Magic City

A small boy with a book, high up in a tree. When I was eight years old somebody gave me *The Magic City* by Edith Nesbit. Nesbit wrote a number of other children's books, which are more famous and better written. But this was the one which I loved and have never forgotten. I did not at the age of eight read deep meanings into it, but I knew that it was somehow special. The story has a coherent architectural plan, covered with a surface frosting of crazy logic. *The Wizard of Oz* was the other book that I used to read over and over again. It has the same qualities. An eight-year-old already has a feeling for such things, even if he spends most of his waking hours climbing trees. *The Magic City* is not just a story about some crazy kids. It is a story about a crazy universe. What I see now, and did not see as an eight-year-old, is that Nesbit's crazy universe bears a strong resemblance to the one we live in.

Edith Nesbit was from every point of view a remarkable woman. Born in 1858, she was intimate with the family of Karl Marx and became a revolutionary socialist long before this was fashionable. She supported herself by writing and brought up a large family of children of mixed parentage. She soon discovered that her survival depended upon her ability to write splendidly bourgeois stories for the children of the rich. Her books sold well, and she survived. She made some compromises with Victorian respectability, but did not lose her inner fire. She wrote *The Magic City* in 1910, when she was fifty-two. By that time her personal struggles were over and she could view the world with a certain philosophic calm.

There are three themes in *The Magic City*. The first is the main theme. The hero is an orphan called Philip who is left alone in a big house and builds a toy city out of the ambient Victorian bric-a-brac. One night he suddenly finds his city grown to full size, inhabited by full-size mythical people and animals, and himself obliged to live in it. After escaping from the city, he wanders through the surrounding country, where every toy house or castle that he ever built is faithfully enlarged and preserved. The book records his adventures as he stumbles through this world of blown-up products of his own imaginings.

The second theme is concerned explicitly with technology. It is a law of life in the magic city that if you wish for anything you can have it. But with this law goes a special rule about machines. If anyone wishes for a piece of machinery, he is compelled to keep it and go on using it for the rest of his life. Philip fortunately escapes from the operation of this rule when he has the choice of wishing for a horse or a bicycle and chooses the horse.

The third theme of the book is the existence of certain ancient prophecies foretelling the appearance of a Deliverer and a Destroyer. Various evil forces are at large in the land, and it is the destiny of the Deliverer to overcome them. But it is also foreordained that a Destroyer will come to oppose the Deliverer and give aid to the forces of darkness. At the beginning Philip is suspected of being the Destroyer. He is only able to vindicate himself by a succession of increasingly noble deeds, which ultimately result in his being acclaimed as the Deliverer. Meanwhile the Destroyer is unmasked and turns out to be the children's nursemaid, a woman of the lower classes whom Philip has always hated. Only once, at the end of the book, Nesbit steps out of character and shows where her real sympathies lie. "I'll speak my mind if I die for it," says the Destroyer as she stands awaiting sentence, "You don't understand. You've never been a servant, to see other people get all the fat and you all the bones. What you think it's like to know if you'd just been born in a gentleman's mansion instead of in a model workman's dwelling you'd have been brought up as a young lady and had the openwork silk stockings?" Even an eight-year-old understands at this point that Philip's heroic virtue is phony and the nursemaid's heroic defiance is real. In an unjust world, the roles of Deliverer and Destroyer become ambiguous. "Think not that I am come to send peace on earth," said Jesus. "I come not to send peace, but a sword."

I do not know how far Nesbit consciously intended *The Magic City* to be an allegory of the human condition. It was only after I descended from the trees, and tasted the joys and sorrows of becoming a scientist, that I began to meditate upon the magic city and to see in it a mirror image of the big world that I was entering. I was plunged into the big world abruptly, like Philip. The big world, wherever I looked, was full of human tragedy. I came upon the scene and found myself playing roles that were half serious and half preposterous. And that is the way it has continued ever since.

I am trying in this book to describe to people who are not scientists the way the human situation looks to somebody who is a scientist. Partly I shall be describing how science looks from the inside. Partly I shall be discussing the future of technology. Partly I shall be struggling with the ethical problems of war and peace, freedom and responsibility, hope and despair, as these are affected by science. These are all parts of a picture which must be seen as a whole in order to be understood. It makes no sense to me to separate science from technology, technology from ethics, or ethics from religion. I am talking here to unscientific people who ultimately have the responsibility for guiding the growth of science and technology into creative rather than destructive directions. If you, unscientific people, are to succeed in this task, you must understand the nature of the beast you are trying to control. This book is intended to help you to understand. If you find it merely amusing or bewildering, it has failed in its purpose. But if you find none of it amusing or bewildering, it has failed even more completely. It is characteristic of all deep human problems that they are not to be approached without some humor and some bewilderment. Science is no exception.

My colleagues in the social sciences talk a great deal about methodology. I prefer to call it style. The methodology of this book is literary rather than analytical. For insight into human affairs I turn to stories and poems rather than to sociology. This is the result of my upbringing and background. I am not able to make use of the wisdom of the sociologists because I do not speak their language. When I see scientists becoming involved in public affairs and trying to use their technical knowledge politically for the betterment of mankind, I remember the words of Milton the poet: "I cannot praise a fugitive and cloistered virtue, unexercised and unbreathed, that never sallies out and sees her adversary." These words, written three hundred years ago, still stand as a monument of human experience, hope and tragedy. They reverberate with echoes of Milton's poetry, his fight for the freedom of the press, his long years of service to the cause of rebellion against monarchy, his blindness, his political downfall, and his final redemption in the writing of *Paradise Lost.* What more can one say that is not by comparison cheap and shallow? We are scientists second, and human beings first. We become politically involved because knowledge implies responsibility. We fight as best we can for what we believe to be right. Often, like Milton, we fail. What more can one say?

A substantial part of this book is autobiographical. I make no apology for that. It is not that I consider my own life particularly significant or interesting to anybody besides myself. I write about my own experiences because I do not know so much about anyone else's. Almost any scientist of my generation could tell a similar story. The important thing, to my mind, is that the great human problems are problems of the individual and not of the mass. To understand the nature of science and of its interaction with society, one must examine the individual scientist and how he confronts the world around him. The best way to approach the ethical problems associated with science is to study real dilemmas faced by real scientists. Since firsthand evidence is the most reliable, I begin by writing about things that happened to me personally. This is another effect of the same individualistic bias that leads me to listen to poets more than to economists.

But I still have to finish what I was saying about *The Magic City* and its three themes. That we live in a world of overgrown toys is too obvious to need explaining. Nikolaus Otto plays for a few years with a toy gasoline engine and—bingo!—we all find ourselves driving cars. Wallace Carothers gets interested in condensation polymers and—zing!—every working-class girl is wearing nylon stockings that are as fancy as the openwork silk that was for Nesbit in 1910 the hated symbol of upper-class privilege. Otto Hahn and Fritz Strassmann amuse themselves with analytical radiochemistry and—boom!—a hundred thousand people in Hiroshima are dead. The same examples also illustrate Nesbit's rule about the consequences of wishing for machinery. Once you have wished for cars, nylons or nuclear weapons, you are stuck with them in a very permanent fashion. But there is one great difference between Philip's world and ours. In his world, every toy castle that he had ever built appeared enlarged. In our world, thousands of scientists play with millions of toys, but only a few of their toys grow big. The majority of technological ventures remain toys, of interest only to specialists and historians. A small number succeed spectacularly and become part of the fabric of our lives. Even with the advantage of hindsight it is difficult to understand why one technology is overwhelmingly successful and another is stillborn. Subtle differences of quality have decisive effects. Sometimes an accident that nobody could have predicted makes a particular toy grow monstrous. When Otto Hahn stumbled upon the discovery of nuclear fission in 1938 he had no inkling of nuclear weapons, no premonition that he was treading on dangerous ground. When the news of Hiroshima came to him seven years later, he was overcome with such grief that his friends were afraid he would kill himself.

Science and technology, like all original creations of the human spirit, are unpredictable. If we had a reliable way to label our toys good and bad, it would be easy to regulate technology wisely. But we can rarely see far enough ahead to know which road leads to damnation. Whoever concerns himself with big technology, either to push it forward or to stop it, is gambling in human lives.

Scientists are not the only people who play with intellectual toys that suddenly explode and cause the crash of empires. Philosophers, prophets and poets do it too. In the long run, the technological means that scientists place in our hands may be less important than the ideological ends to which these means are harnessed. Technology is powerful but it does not rule the world. Nesbit lived long enough to see one tenth of mankind ruled by ideas that the man known in the family as "Old Nick" had worked out in his long quiet days at the British Museum. Old Nick, alias Karl Marx, was the father-in-law of her friend Edward Aveling.

Marx was in his own lifetime a larger-than-life figure, and after his death became Deliverer to half the world and Destroyer to the other half. There is a deep-rooted tendency in the human soul that builds myths of Deliverers and Destroyers. These myths, like other myths, have a foundation in truth. The world of science and technology may appear on the surface to be rational, but it is not immune to such myths. The great figures of science have a quality, an intensity of will and character, that sets them apart from ordinary scientists as Marx stands apart from ordinary economists. We shall not understand the dynamics of science and technology, just as we shall not understand the dynamics of political ideology, if we ignore the dominating influence of myths and symbols.

I was lucky to hear the economist John Maynard Keynes, a few years before his death, give a lecture about the physicist Isaac Newton. Keynes was at that time himself a legendary figure, gravely ill and carrying a heavy responsibility as economic adviser to Winston Churchill. He had snatched a few hours from his official duties to pursue his hobby of studying Newton's unpublished manuscripts. Newton had kept his early writings hidden away until the end of his life in a big box, where they remained until quite recently. Keynes was speaking in the same old building where Newton had lived and worked 270 years earlier. In an ancient, dark, cold room, draped with wartime blackout curtains, a small audience crowded around the patch of light under which the exhausted figure of Keynes was huddled. He spoke with passionate intensity, made even more impressive by the pallor of his face and the gloom of the surroundings. Here are some extracts from his talk.

As one broods over these queer collections, it seems easier to understand—with an understanding which is not, I hope, distorted in the other direction—this strange spirit, who was tempted by the Devil to believe, at the time when within these walls he was solving so much, that he could reach *all* the secrets of God and Nature by the pure power of mind—Copernicus and Faustus in one.

A large section, judging by the handwriting among the earliest, relates to alchemy—transmutation, the philosopher's stone, the elixir of life.

All his unpublished works on esoteric and theological matters are marked by careful learning, accurate method and extreme sobriety of statement. They are just as *sane* as the *Principia,* if their whole matter and purpose were not magical.

Why do I call him a magician? Because he looked on the whole universe and all that is in it *as a riddle,* as a secret which could be read by applying pure thought to certain evidence, certain mystic clues which God had laid about the world to allow a sort of philosopher's treasure hunt to the esoteric brotherhood. . . . He *did* read the riddle of the heavens. And he believed that by the same powers of his introspective imagination he would read the riddle of the Godhead, the riddle of past and future events divinely foreordained, the riddle of the elements and their constitution from an original undifferentiated first matter, the riddle of health and of immortality.

Newton was admittedly an extreme case. When I quote these words of Keynes I do not mean to imply that every great scientist should devote half his time to magical mumbo-jumbo. I am suggesting that anyone who is transcendentally great as a scientist is likely also to have personal qualities that ordinary people would consider in some sense superhuman. If he were not gifted with extraordinary strength of character, he could not do what he does in science. Thus it is not surprising that traditional mythology links the figure of the scientist with that of the Magus. The Magi were the priests of the ancient Zoroastrian religion of Persia, and the word "magic" is derived from their name. The myth of the scientist-Magus appears in its most complete form in the legend of Faust, the learned man who sells his soul to the Devil in return for occult knowledge and magical power. The remarkable thing about the Faust legend is that everybody to some extent still believes in it. When you say that some piece of technology is a Faustian bargain, everybody knows what you mean. Somewhere below the level of rational argument, the myth is alive.

I shall talk later about various scientists who have acquired public reputations as deliverers or destroyers. Such reputations are often transient or even fraudulent, but they are not meaningless. They indicate a recognition by the public that somebody has done something that matters. The public also recognizes a special personal quality in these people. The greatest and most genuine deliverer in my lifetime was Einstein. His special quality was universally recognized, although it is not easy to describe in words. I shall not talk about Einstein since I did not know him personally and I have nothing to add to what has already been said by others.

In the magic city there are not only deliverers and destroyers but also a great multitude of honest craftsmen, artisans and scribes. Much of the joy of science is the joy of solid work done by skilled workmen. Many of us are happy to spend our lives in collaborative efforts where to be reliable is more important than to be original. There is a great satisfaction in building good tools for other people to use. We do not all have the talent or the ambition to become prima donnas. The essential factor which keeps the scientific enterprise healthy is a shared respect for quality. Everybody can take pride in the quality of his own work, and we expect rough treatment from our colleagues whenever we produce something shoddy. The knowledge that quality counts makes even routine tasks rewarding.

Recently a new magus has appeared upon the scene: a writer, Robert Pirsig, with a book, *Zen and the Art of Motorcycle Maintenance*. His book explores the dual nature of science, on the one hand science as dedicated craftsmanship, on the other hand science as intellectual obsession. He dances with wonderful agility between these two levels of experience. On the practical level, he describes for unscientific readers the virtue of a technology based upon respect for quality. The motorcycle serves as a concrete example to illustrate the principles which should govern the practical use of science. On the intellectual level, Pirsig weaves into the discussion of technology a narrative of his own quest for philosophical understanding, ending with a mental collapse and reintegration. Phaedrus, the alter ego of Pirsig, is a spirit so dominated by intellectual struggle that he has become insane. In order to survive as a human being, Pirsig has driven Phaedrus out of his consciousness, but Phaedrus comes back to haunt him. The small boy Chris who rides on the back of the motorcycle succeeds in the end in bringing Phaedrus and Pirsig together. In a strange fashion, this personal drama adds insight to Pirsig's vision of technology. Pirsig is by profession a writer and not a scientist. But he has struggled to order rationally the whole of human experience, as Newton struggled three hundred years earlier. He has pored over the pre-Socratic Greek philosophers in his study in Montana, as Newton pored over the ancient alchemical texts in his laboratory in Cambridge. The struggle brought both of them to the edge of madness. Each of them in the end abandoned the greater part of his design and settled for a more limited area of understanding. But Pirsig's message to our generation, as we try to come to terms with technology, is deepened and strengthened because he is who he is and has seen what he has seen:

> The magus Zoroaster, my dead child,
> Met his own image walking in the garden.
> That apparition, sole of men, he saw.

2

The Redemption of Faust

A year before the beginning of the Second World War, I got hold of a copy of Piaggio's *Differential Equations*. This did not come from my teachers. At that time I had never been near a university or a technical library. My door to knowledge was a little handwritten letter which I sent to various book publishers: "Dear Sirs, Please would you send to the above address a catalog of your scientific publications. Yours faithfully." Sure enough, within a few days the catalog would arrive. The most exciting catalogs came from the Cambridge University Press. They had long lists of books resulting from the *Challenger* expedition of 1872–76. The voyage of H.M.S. *Challenger* was the first worldwide scientific exploration of the oceans, and that one little ship brought back such a wealth of material that they were still selling books about it in 1938. I wondered vaguely whether there might not one day be another such voyage, and whether I might not have a chance to sail on it. But the *Challenger* volumes were far too expensive for me to buy, and so my career as an oceanographer ended before it began.

Mathematics was cheaper. I had read some of the popular literature about Einstein and relativity, and had found it very unsatisfying. Always when I thought I was getting close to the heart of the matter, the author would say, "But if you really want to understand Einstein you have to understand differential equations," or words to that effect. I did not have a clear idea of what a differential equation was, but I knew it was Einstein's language and I had to learn it. So it was a day of great joy when a skimpy catalog arrived from G. Bell and Sons Limited, containing the item *Differential Equations,* by H. T.

H. Piaggio, twelve shillings and sixpence. I had never heard of Piaggio, but twelve and six was within my range, and I went at once to the bookshop to put in my order. In due course the book arrived, rather small and modestly bound in light-blue cloth. I was too busy during the school term to give my attention to it, so I saved it for the Christmas vacation.

My school vacations were mostly spent at a cottage on the shore which my father had bought as a holiday home. He was a musician. He worked for many years as music teacher in the same school which I attended as a boy in Winchester. He enjoyed the life of a school-teacher, with three months vacation a year and plenty of time left free for conducting and composing even during the school terms. His best-known work is "The Canterbury Pilgrims," a setting of the Prologue of Chaucer's *Canterbury Tales* for solo voices, chorus and orchestra. It was first performed at Winchester when I was seven years old. It is dedicated "to M.L.D., who prepared the words." That is my mother, who shared with him an intense affection for Chaucer and for the characters that Chaucer immortalized. We often encountered modern reincarnations of one or another of Chaucer's pilgrims. Then my parents would exchange glances, my mother would whisper a line of Chaucer, or my father would quietly hum the appropriate tune. The well-fed clergy of Winchester would remind them of Chaucer's Monk:

> He was a lord full fat and in great point;
> His eyes were bright and rolling in his head,
> That gleamed like a fire beneath a pot.

A doctor driving a Rolls-Royce along our street would suggest Chaucer's Doctor of Physic:

> He kept all that he won in pestilence.
> For gold in physic is a cordial,
> Therefore he loved gold in special.

The sights and sounds of the English countryside would call to mind Chaucer's descriptions of it:

> And small birds make melody
> That sleep all night with open eye,
> So worketh nature in their hearts.

In vacation time, when we were at the cottage, my father regularly composed for three hours every morning. In the afternoons he loved to potter around and improve his forty acres of waterlogged land. The land could do with a great deal of improving, since it lay below sea level on the south coast of England and had been repeatedly flooded with salt water. We were supposed to maintain our section of the dike which kept the sea out. The land was drained by a system of ditches which flowed into bunnies. A bunny was a pipe laid under the dike, with a wooden clapper which opened to let water out from land to sea at low tide and closed to keep the sea from coming in at high tide. The bunnies were my father's pride and joy. He was never happier than when he was standing waist deep in cold black mud to excavate a clogged bunny. When the bunnies were working smoothly he would excavate the ditches. Only one thing was missing. To make his happiness complete he would have liked to have his growing son out there with him in the mud to give him help and companionship.

My idea of a joyful Christmas vacation was different. I arrived at the cottage on the coast with my precious Piaggio and did not intend to be parted from him. I soon discovered that Piaggio's book was ideally suited to a solitary student. It was a serious book, and went rapidly enough ahead into advanced territory. But unlike most advanced texts, it was liberally sprinkled with "Examples for Solution." There were more than seven hundred of these problems. The difference between a text without problems and a text with problems is like the difference between learning to read a language and learning to speak it. I intended to speak the language of Einstein, and so I worked my way through the problems. I started at six in the morning and stopped at ten in the evening, with short breaks for meals. I averaged fourteen hours a day. Never have I enjoyed a vacation more.

After a while my parents became worried. My mother looked sadly at me and quoted from Chaucer's Clerk of Oxenford:

> Of study took he most care and most heed,
> Not a word spake he more than was need.

She warned me that I would ruin my health and burn out my brains if I went on like this. My father begged me, just for a few hours, to stop calculating and help him with his ditches. But their entreaties only made me more stubborn. I was in love with mathematics, and nothing else mattered. I was also acutely aware of the approaching war. We did not then know that it was our last peacetime Christmas, but we could all see the war coming. I knew what had happened to the English boys who were fifteen at the start of the First World War and arrived in the trenches in 1917 and 1918. In all probability I had not many years to live, and every hour spent not doing mathematics was a tragic waste. How could my father be so blind as to wish to ruin my few remaining days on earth with his dull ditches? I looked on his blindness more in sorrow than in anger.

In those days my head was full of the romantic prose of E. T. Bell's book *Men of Mathematics,* a collection of biographies of the great mathematicians. This is a splendid book for a young boy to read (unfortunately, there is not much in it to inspire a girl, with Sonya Kowalewska allotted only half a chapter), and it has awoken many people of my generation to the beauties of mathematics. The most memorable chapter is called "Genius and Stupidity" and describes the life and death of the French mathematician Galois, who was killed in a duel at the age of twenty. In spite of all the sentimental mush that has been written about him, he was a genuine genius and his death was a genuine tragedy. Galois groups and Galois fields are still after 140 years a living part of mathematics. E. T. Bell describes the last night before the fatal duel: "All night he had spent the fleeting hours feverishly dashing off his scientific last will and testament, writing against time to glean a few of the great things in his teeming mind before the death which he foresaw could overtake him. Time after time he broke off to scribble in the margin 'I have not time; I have not time,' and passed on to the next frantically scrawled outline. What he wrote in those desperate last hours before the dawn will keep generations of mathematicians busy for hundreds of years. He had found, once and for all, the true solution of a riddle which had tormented mathematicians for centuries: under what conditions can an equation be solved?" These words added a touch of noble pathos to the long hours that I was spending with Piaggio. If I was destined to die at the age of nineteen, like so many of the junior officers of the First World War, then I would have one year less than Galois.

Our Christmas vacation lasted a full month. Before it was over I was coming near to the end of Piaggio's seven hundred examples. I began to skip a few of them. I was even willing to set aside an hour or two to take a walk with my mother. My mother had been waiting a long time for a chance to talk to me. She was well prepared. So a few days before the end of vacation we went out together.

My mother was a lawyer by profession and intensely interested in people. She loved the Latin and Greek poets. She began her lecture with a quotation from the play *The Self-Tormentor* by the African slave Terentius Afer, who became the greatest Latin playwright: *"Homo sum: humani nil a me alienum puto."* "I am human and I let nothing human be alien to me." This was the creed by which she lived a long and full life until she died at the age of ninety-four. She told me then, as we walked along the dike between the mud and the open water, that this should also be my creed. She understood my impatience, and my passion for the abstract beauties of Piaggio. But she begged me not to lose my humanity in my haste to become a mathematician. You will regret it deeply, she said, when one day you are a great scientist and you wake up to find that you have never had time to make friends. What good will it do you to prove the Riemann hypothesis, if you have no wife and no children to share your triumph? You will find even mathematics itself will grow stale and bitter if that is the only thing you are interested in.

I listened to all this carelessly, knowing that I had no use for it yet but could come back to it later. After my mother had finished with Terence the African, she began again with Goethe's *Faust*. She told me the story of Faust from Goethe's First Part. How Faust works day and night at his books, consumed by the ambition to know everything and command the forces of nature. How he becomes more and more self-centered and more and more dissatisfied. How he goes altogether to the bad and loses his soul to the Devil in exchange for knowledge and power. How his attempt to find happiness with Gretchen leads only to misery and tragedy, since he is incapable of unselfish love and can only compel her to love him on his own terms. Some years later when the film *Citizen Kane* came over from America and I went to see it, I suddenly found myself in tears and realized it was because Orson Welles's artistry made my mother's image of Faust come alive again. Kane and Faust, Faust and Kane and I, each of us damned to eternal friendlessness by our selfish ambitions.

But my mother did not leave me comfortless. She went on to talk at length about *Faust* Part Two, the work of Goethe's old age, in which Faust is finally redeemed. It is agreed between the heavenly and infernal powers that Faust will be redeemed if he can ever find a moment of true happiness in which his soul is at peace with itself. Through many tedious pages of verse he searches in vain for the blissful moment. He meets with Helen of Troy and various other mythological personages, tries his hand as a general in command of an army, but finds no satisfaction in it. In the end, when he is old and blind, he comes to a Dutch village where the whole population is engaged in a desperate struggle to defend their land against the sea. The people of the village are out at the dike, digging and pumping, working together with all their might against the common danger. Faust joins them and throws himself into the work without a thought for his frail condition. Suddenly he realizes that this is the blissful moment that he has been seeking all his life, the joy of working together with his fellow men in a common endeavor, the joy of being immersed in a cause larger than himself. So he dies redeemed and is carried off to heaven by an angelic choir. Afterward when I happened once to read the closing pages of *Faust* Part Two, I was surprised to find that this vividly remembered scene of the Dutch villagers at the dike owes more to my mother's imagination than to Goethe. What Goethe wrote is only a pale shadow of it. It is a pity Goethe never heard her version of the story.

So my road to redemption was clear. Down to the ditches with my father. Grudgingly, I joined him in the mud for one afternoon. No angels came to waft me to heaven.

After the vacation was over, I went back to school, quickly finished Piaggio and was ready to begin on Einstein. Unfortunately, none of my book catalogs offered anything written by Einstein, and for a while I was stuck. I ordered from the Cambridge University Press Eddington's *Mathematical Theory of Relativity* and made do with that. After Piaggio it went quite easily. Meanwhile my mother's words of wisdom were slowly sinking into the subconscious levels of my mind and preparing fresh surprises for me. I agreed with her in theory when she said that human solidarity and companionship were the essential ingredients of a satisfactory life. But in practice, for the time being, I saw little that I could do about it.

Like everybody else at that time, I worried a great deal about the approaching war. I was not concerned about winning it or losing it. It seemed then that there was equally small chance that anything worth preserving would survive the war, whether we won it or lost it. The war was for me an unconditional evil. I was concerned only to do whatever I could to stop it from beginning. And the only way to stop it was to change the hearts and minds of the warmakers on both sides. It was clear that only a radical change in their way of thinking could do the job.

I tried hard to understand the deeper causes of the hatreds that were driving us to war. I concluded that the basic cause of war was injustice. If all men had a fair share of the world's goods, if all of us were given an equal chance in the game of life, then there would be no hatred and no war. So I asked myself the age-old questions, why does God permit war, and why does God permit injustice, and I found no answers. The problem of injustice seemed to me even more intractable than the problem of war. I was gifted with brains, good health, books, education, a loving family, not to mention food, clothing and shelter. How could I imagine a world in which the Welsh coal miner's son and the Indian peasant would be as lucky as I was?

Enlightenment came to me suddenly and unexpectedly one afternoon in March when I was walking up to the school notice board to see whether my name was on the list for tomorrow's football game. I was not on the list. And in a blinding flash of inner light I saw the answer to both my problems, the problem of war and the problem of injustice. The answer was amazingly simple. I called it Cosmic Unity. Cosmic Unity said: There is only one of us. We are all the same person. I am you and I am Winston Churchill and Hitler and Gandhi and everybody. There is no problem of injustice because your sufferings are also mine. There will be no problem of war as soon as you understand that in killing me you are only killing yourself.

For some days I quietly worked out in my own mind the metaphysics of Cosmic Unity. The more I thought about it, the more convinced I became that it was the living truth. It was logically incontrovertible. It provided for the first time a firm foundation for ethics. It offered mankind the radical change of heart and mind that was our only hope of peace at a time of desperate danger. Only one small problem remained. I must find a way to convert the world to my way of thinking.

The work of conversion began slowly. I am not a good preacher. After I had expounded the new faith two or three times to my friends at school, I found it difficult to hold their attention. They were not anxious to hear more about it. They had a tendency to run away when they saw me coming. They were good-natured boys, and generally tolerant of eccentricity, but they were repelled by my tone of moral earnestness. When I preached at them I sounded too much like the headmaster. So in the end I made only two converts, one whole-hearted and one half-hearted. Even the whole-hearted convert did not share in the work of preaching. He liked to keep his beliefs to himself. I, too, began to suspect that I lacked some of the essential qualities of a religious leader. Relativity was more in my line. After a few months I gave up trying to make converts. When some friend would come up to me and say cheerfully, "How's cosmajoonity doing today?" I would just answer, "Fine, thank you," and let it go at that.

In the summer vacation I made one last attempt at a conversion. I asked my mother to come out for another walk along the dike and I laid before her my message of hope and glory. She was obviously very happy to see that I had discovered there are more things in heaven and earth than differential equations. She smiled at me and said very little. After I had finished talking I asked her what she thought about it all. She answered slowly, "Yes. I have believed something rather like that for a very long time."

3

The Children's Crusade

Wing Commander MacGown was chief medical officer in the Pathfinder Force of the Royal Air Force Bomber Command. He was on Lancaster 83Q, taking off for Berlin from Wyton Air Force Base, at a very desperate time in January 1944. Wyton was the home of 83 Squadron, one of the original pathfinder squadrons which had been leading the night attacks on German cities since the pathfinders began. I stood by the runway, facing into a cold wet wind, and watched the twenty Lancasters of 83 Squadron take off into blackness. They were heavily overloaded and took a long time to get airborne. The Lancaster had a phenomenal capacity for carrying bombs. The permissible overload had been raised several times since Lancasters began operations in 1942. After the bombers took off I went inside for a cup of tea.

Wyton was as ugly as a wartime military base can be. Endless puddles, barracks, warehouses full of bombs, rusting wreckage of damaged equipment not worth repairing. For two months 83 Squadron had been going out night after night, whenever the weather was not completely impossible, to bomb Berlin. On the average they were losing an aircraft each time they went out. Each Lancaster carried a crew of seven.

Bomber Command was putting its maximum effort into the repeated attacks on Berlin that winter, because it was the last chance to do decisive damage to the German war economy before the Western armies would begin the invasion of Europe. The boys who flew in the Lancasters were told that this battle of Berlin was one of the decisive battles of the war and that they were winning it. I did not know how many of them believed what they were told. I knew only that what they were told was untrue. By January 1944 the battle was lost. I had seen the bomb patterns, which showed bombs scattered over an enormous area. The bomber losses were rising sharply. There was no chance that our continuing the offensive in this style could have any decisive effect on the war. It was true that Berlin contained a great variety of important war industries and administrative centers. But Bomber Command was not attempting to find and attack these objectives individually. We merely showered incendiary bombs over the city in as concentrated a fashion as possible, with a small fraction of high-explosive bombs to discourage the fire-fighters. Against this sort of attack the defense could afford to be selective. Important factories were protected by fire-fighting teams who could deal quickly with incendiaries falling in vital areas. Civilian housing and shops could be left to burn. So it often happened that Bomber Command "destroyed" a city, and photographic reconnaissance a few weeks later showed factories producing as usual amid the rubble of burnt homes.

On just two occasions during the war, a Bomber Command incendiary attack was outstandingly successful. This happened first in Hamburg in July 1943. We started so many fires in a heavily built-up area that a fire storm developed, a hurricane of flame that killed forty thousand people and destroyed everything in its path. None of our other attacks had produced effects that were a tenth as destructive as the effects of a fire storm. The only way we could have won a militarily meaningful victory in the battle of Berlin was to raise a fire storm there. Conceivably, a giant fire storm raging through Berlin could have fulfilled the dreams of the men who created Bomber Command. "Victory through Air Power" was their slogan. But I knew in January 1944 that this was not going to happen. A fire storm could happen only when the bombers were able to bomb exceptionally accurately and without serious interference from the defenses. Under our repeated battering the defenses of Berlin were getting stronger, and the scatter of the bombing was getting worse. Only once more, a year after my visit to Wyton, when Germany was invaded and almost overrun, we succeeded again in raising a fire storm. That was in February 1945, in Dresden.

I was a civilian scientist working at Bomber Command headquarters. I had come a long way since the innocent days of Cosmic Unity.

I belonged to a group called the Operational Research Section, which gave scientific advice to the commander in chief. I was engaged in a statistical study to find out whether there was any correlation between the experience of a crew and their chance of being shot down. The belief of the Command, incessantly drummed into the crews during their training and impressed on the public by the official propaganda machine, was that a crew's chance of surviving a mission increased with experience. Once you get through the first five or ten missions, the crews were told, you will know the ropes and you will learn to spot the German night fighters sooner and you will stand a much better chance of coming home alive. To believe this was undoubtedly good for the boys' morale. Squadron commanders, all of them survivors of many missions, sincerely believed that they owed their survival to their personal qualities of skill and determination rather than to pure chance. They were probably right. It had been true in the early years of the war that experienced crews survived better. Before I arrived at Bomber Command, the Operational Research Section had made a study which confirmed the official doctrine of survival through experience. The results of that study had been warmly accepted by everybody.

Unfortunately, when I repeated the study with better statistics and more recent data, I found that things had changed. My analysis was based on complete records and carefully excluded any spurious correlations caused by the fact that inexperienced crews were often given easier missions. My conclusion was unambiguous: the decrease of loss rate with experience which existed in 1942 had ceased to exist in 1944. There were still many individual cases of experienced crews by heroic efforts bringing home bombers so badly damaged that a novice crew in the same situation would almost certainly have been lost. Such cases did not alter the fact that the total effect of all the skill and dedication of the experienced crews was statistically undetectable. Experienced and inexperienced crews were mown down as impartially as the boys who walked into the German machine gun nests at the battle of the Somme in 1916.

The disappearance of the correlation between experience and loss rate ought to have been recognized by our commander in chief as a warning signal, telling him that he was up against something new. In the Operational Research Section we had a theory to explain why experience no longer saved bombers. We now know that our theory was correct. The theory was called "Upward-Firing Guns." Each bomber had four crew members constantly searching the sky for fighters, the pilot and bomb aimer in front and the two gunners in the tail and mid-upper gun turrets. Vertically underneath the bomber was a blind spot. Conventionally armed fighters would not have been able to approach the bomber from underneath and shoot it down without being seen. But increasing numbers of the German fighters were not conventionally armed. They had cannon pointing vertically upward, with a simple periscope gun sight arranged so that the pilot could take careful aim as he flew quietly below the bomber. The main problem for the fighter pilot was to avoid being hit by any large pieces as the bomber disintegrated.

83 Squadron, being an old pathfinder squadron, had more than its share of experienced crews. The normal tour of duty for a crew in a regular squadron was thirty missions. The loss rate during the middle years of the war averaged about four percent. This meant that a crewman had three chances in ten of completing a normal tour. The pathfinder crews signed on for a double tour of sixty missions. They had about one chance in eleven of completing the double tour. During the winter of 1943–44, with the repeated attacks on Berlin, the losses were higher than average and the chances of survival smaller.

I had come to Wyton from Command headquarters to see how various radar countermeasures against fighters were working. The radars worked all right, but they were not much use because they could not distinguish fighters from bombers. I also hoped to pick up information at Wyton that would be helpful for my study of the effects of experience on loss rates. I thought I might talk with some of the experienced crews, gather firsthand impressions, and get a feeling for what was really happening in the nightly battles over Berlin. But it soon became clear that serious conversations between crews and civilian outsiders were impossible. Above all, the subject of survival rates was taboo. The whole weight of Air Force tradition and authority was designed to discourage the individual airman from figuring the odds. Airmen who thought too much about the odds were likely to crack up. Airmen who talked about such matters to their crewmates were a danger to the discipline of the squadron. Stringent precautions were taken to ensure that any of our Command headquarters documents that discussed survival rates should not reach the squadrons. In the squadrons the old rule "Theirs not to reason why, Theirs but to do and die" was still in force.

The crewmen were not forbidden to talk to me. They could talk as much as they liked. But what could they say to me, or what could I say to them, across the gulf that separated us? They were mostly twenty-year-old boys, the same age as I. They had faced flaming death thirty times and would face it thirty times more if they were lucky. I had not, and would not. They knew, and I knew that they knew, that I was one of those college-educated kids who found themselves cushy civilian jobs and kept out of harm's way. How could two twenty-year-olds, separated by such a barrier, talk to each other about anything important?

The one person at Wyton to whom I could talk freely was Wing Commander MacGown. He was responsible for the mental as well as physical health of the crews of the eight pathfinder squadrons. A tall, white-haired officer, he seemed to me very old although he cannot have been much over forty. He was the ultimate authority who decided, when one of the boys began to show signs of mental crack-up, whether he should be kept on operations or transferred out of the squadron. There was no easy way out for boys who cracked. The rules of the Command were designed to ensure that crewmen should consider transfer a fate worse than death. When a boy was transferred for mental reasons, the cause of transfer was officially recorded as "Lack of Moral Fibre." He was, in effect, officially declared to be a coward and thereafter assigned to menial and humiliating duties. In spite of the public disgrace and dishonor that they had to endure, the number who cracked was not small. At Command headquarters, we knew that the number transferred out of squadrons before the end of their tour was roughly equal to the number completing the full tour. We were not allowed to know how many of those transferred were mental cases. But Wing Commander Mac-Gown knew.

I was astonished, at our first meeting, when MacGown told me he was flying to Berlin that night. He said the crews loved to have him go along with them. It was well known in the squadron that the plane with the Doc on board always came home safely. He had already been to Berlin and back six times in the last two months. At first I thought he must be crazy. Why should an elderly doctor with a full-time staff job risk his life repeatedly on these desperately dangerous missions? Afterward I understood. It was the only way he could show these boys for whose bodies and souls he was responsible that he really cared for them. It was the only way he could face the boys who cracked and declare them "lacking in moral fibre" without losing his own self-respect.

While MacGown and twenty times seven crewmen were on their way to Berlin, there was a beer party for the spare crews who for one reason or another were not needed on this operation. The boys drank a great deal of beer and sang their squadron songs.

> We take our bombs to Germany,
> We don't bring them back—

they sang, and at the end of each verse the refrain

> Eighty-three squadron—
> Eighty-three men.

It was the saddest beer party I ever attended. Early in the morning we heard the Lancasters coming home. Only one was missing. It was not MacGown's.

After my visit to Wyton, I decided that the only honorable thing to do was to quit my job at Command headquarters and enlist as a crewman. Because of my mathematical training I expected they would accept me as a navigator. But before taking any such drastic action I discussed the whole situation with my mother. My mother understood at once what was at stake. She saw that it would be useless to appeal directly to my cowardice. Instead she appealed to my incompetence. "You would be absolutely hopeless as a navigator," she said. "You would get lost every time. Of course I won't argue against your going and getting yourself killed if you think that is the right thing to do. But it would be a terrible waste of an airplane." Her words had the desired effect. I gave up the idea of heroic self-sacrifice and went quietly back to work at Bomber Command.

During that winter, while we were attacking Berlin, the Germans used to send a few bombers over London from time to time. The German attacks were on a minuscule scale compared with ours, and they cannot have had any other purpose than to boost the morale of the Berliners. We had carried out similar token raids on Berlin in 1940 when London was under serious attack. So when the German planes came droning overhead in February 1944 I stayed in bed and did not bother to go down to the cellar. I thought of the German boys up there, risking their lives to provide morning copy for the writers in the Propaganda Ministry. I was meditating upon the overwhelming irrelevance of this game of tit-for-tat bombing to the serious war that we were supposed to be engaged in. Just then came a shattering explosion and my bedroom windows lay in splinters on the floor. The Institut Français, two houses away on the corner of Queen's Gate and Prince Consort Road, had taken a direct hit. The Institut was the cultural center for the French community in London before the war. It was said that the prewar French had not been happy when de Gaulle came over from France in 1940 and without any legal authorization claimed for himself the leadership of the Free French forces. There had been sporadic feuding between the Institut people and de Gaulle all through the war. My mother and I went out into the street to watch the Institut burn. It made a glorious blaze in the winter night. Perhaps, after all, those boys up there were not German but French, sent by de Gaulle to pay off an old grudge. Whichever way you looked at it, it made no sense.

In the Operational Research Section, those of us who studied the causes of bomber losses thought we had a promising idea for reducing the losses. We wanted to rip the two gun turrets with all the associated machinery and ammunition out of the bombers and reduce the crew from seven to five. The evidence that loss rate did not decrease with experience confirmed our belief that gunners were of little use for defending bombers at night. The basic trouble with the bombers was that they were too slow and too heavily loaded. The gun turrets were heavy and aerodynamically awkward. We estimated that a bomber with turrets ripped out and the holes covered with smooth fairings would fly fifty miles an hour faster and be much more maneuverable. Bomber losses varied dramatically from night to night. We knew that the main cause of the variation was the success or failure of the German fighter controllers in directing the fighters into the bomber stream before it reached the target. An extra fifty miles an hour might have made an enormous difference. At the very least, we urged, the Command could try the experiment of ripping the turrets out of a few squadrons. They would then soon see whether the gunless Lancasters were shot down more or less than the others. Privately, I had another reason for wanting to rip out the turrets. Even if the change did not result in saving a single bomber, it would at least save the lives of the gunners.

All our advice to the commander in chief was channeled through the chief of our section, who was a career civil servant. His guiding principle was only to tell the commander in chief things that the commander in chief liked to hear. His devotion to this principle earned him the expected promotion at the end of the war and led later to the inevitable knighthood. I still remember the shock I felt the first time I saw our chief in action. I happened to be in his office when a WAAF sergeant came in with a bomb plot of a recent attack on Frankfurt. As usual, the impact points deduced from flash photographs were plotted on a map of the city with a three-mile circle drawn around the aiming point. The plot was supposed to go to the commander in chief together with our analysis of the raid. Our chief looked glumly at it for a few seconds and then gave it back to the sergeant. "Awfully few bombs inside the circle," he said. "You'd better change that to a five-mile circle before it goes in." After this experience, I was not surprised to learn that our chief took a dim view of our suggestion that bombers might survive better without gun turrets. This was not the kind of suggestion that the commander in chief liked to hear, and therefore our chief did not like it either. To push the idea of ripping out gun turrets, against the official mythology of the gallant gunner defending his crewmates, and against the massive bureaucratic inertia of the Command, would have involved our chief in a major political battle. Perhaps it was a battle he could not have hoped to win. In any case, the instinct of a career civil servant told him to avoid such battles. The gun turrets remained in the bombers, and the gunners continued to die uselessly until the end of the war.

I shared an office at Command headquarters with a half-Irish boy of my own age called Mike O'Loughlin. He had been a soldier in the army, developed epilepsy, and was given a medical discharge. He knew less mathematics than I did but more about the real world. When we looked around us at the brutalities and stupidities of the Command, I got depressed and Mike got angry. Anger is creative; depression is useless.

One of the things that Mike was angry about was escape hatches. Every bomber had a trap door in the floor through which the crew was supposed to jump when the captain gave the order to bail out.

The official propaganda gave the crews the impression that they had an excellent chance of escaping by parachute if their plane should be so unlucky as to be shot down. They were generally more worried about being lynched by infuriated German civilians than about being trapped in a burning aircraft. In fact, lynching by civilians never happened, and only a small number of airmen were shot by the Gestapo after being captured. A far larger number died because they were inadequately prepared for the job of squeezing through a small hole with a bulky flying suit and parachute harness, in the dark, in a hurry, in an airplane rapidly going out of control. The mechanics of bailing out was another taboo subject which right-thinking crewmen were not encouraged to discuss. The actual fraction of survivors among the crews of shot-down planes was a secret kept from the squadrons even more strictly than the odds against their completing an operational tour. If the boys had found out how small was the fraction who succeeded in bailing out after being hit, some of them might have been tempted to jump too soon.

Mike was no respecter of official taboos. He managed to collect fairly complete information concerning the numbers of crewmen, from missing aircraft of various types, who turned up as prisoners of war. The numbers that he found were startling. From American bombers shot down in daylight, about fifty percent escaped. From the older types of British night bomber, Halifax and Stirling, about twenty-five percent. From Lancasters, fifteen percent. The Lancaster was our newest bomber and in every other respect superior to the Halifax and Stirling. The older bombers were being phased out and the squadrons were being rapidly converted to Lancasters. Mike was the only person in the entire Command who worried about what this would do to the boys who were shot down.

It was easy to argue that the difference in the escape rate between American bombers and Halifaxes and Stirlings was attributable to the difference in circumstances between day and night bombing. The Americans may have had more warning before they were hit and more time to organize their departure. It was obviously easier to find the way out by daylight than in the dark. No such excuses could account for the difference between Halifaxes and Lancasters. Mike discovered quickly the true explanation for the low escape rate from Lancasters. The escape hatch of a Halifax was twenty-four inches wide; the width of a Lancaster hatch was twenty-two inches. The missing two inches probably cost the lives of several thousand boys.

Mike spent two years in a lonely struggle to force the Command to enlarge the Lancaster hatch. Ultimately he succeeded. It was an astonishing triumph of will power over bureaucracy, one epileptic boy overcoming the entrenched inertia of the military establishment. But Mike's progress was maddeningly slow. After he had collected the information on escape rates, it took many months before the Command would officially admit that a problem existed. After the problem had been officially recognized, it took many months to persuade the companies who built the Lancaster that they ought to do something about it. After the companies started to work on the problem, it took many months before a bigger hatch was designed and put into production. The bigger hatch became standard only when the war was almost over and the crews who might have been saved by it were mostly dead. When the total casualty figures for Bomber Command were added up at the end of the war, the results were as follows: Killed on operations, 47,130. Bailed out and survived, 12,790, including 138 who died as prisoners of war. Escape rate, 21.3 percent. I always believed that we could have come close to the American escape rate of fifty percent if our commanders had been seriously concerned about the problem.

We killed altogether about 400,000 Germans, one third of them in the two fire storms in Hamburg and Dresden. The Dresden fire storm was the worst. But from our point of view it was only a fluke. We attacked Berlin sixteen times with the same kind of force that attacked Dresden once. We were trying every time to raise a fire storm. There was nothing special about Dresden except that for once everything worked as we intended. It was like a hole in one in a game of golf. Unfortunately, Dresden had little military importance, and anyway the slaughter came too late to have any serious effect on the war.

Kurt Vonnegut wrote a book called *Slaughterhouse-Five, or The Children's Crusade* about the Dresden raid. For many years I had intended to write a book on the bombing. Now I do not need to write it, because Vonnegut has written it much better than I could. He was in Dresden at the time and saw what happened. His book is not only good literature. It is also truthful. The only inaccuracy that I found in it is that it does not say that the night attack which produced the holocaust was a British affair. The Americans only came the following day to plow over the rubble. Vonnegut, being American, did not want to write his account in such a way that the whole thing could be blamed on the British. Apart from that, everything he says is true. One of the most truthful things in the book is the subtitle, "The Children's Crusade." Vonnegut explains in his introduction how the wife of one of his friends got angry and made him use that subtitle. She was right. A children's crusade is just what the whole bloody shambles was.

Bomber Command might have been invented by some mad sociologist as an example to exhibit as clearly as possible the evil aspects of science and technology: The Lancaster, in itself a magnificent flying machine, made into a death trap for the boys who flew it. A huge organization dedicated to the purpose of burning cities and killing people, and doing the job badly. A bureaucratic accounting system which failed utterly to distinguish between ends and means, measuring the success of squadrons by the number of sorties flown, no matter why, and by the tonnage of bombs dropped, no matter where. Secrecy pervading the hierarchy from top to bottom, not so much directed against the Germans as against the possibility that the failures and falsehoods of the Command should become known either to the political authorities in London or to the boys in the squadrons. A commander in chief who accepted no criticism either from above or from below, never admitted his mistakes, and appeared to be as indifferent to the slaughter of his own airmen as he was to the slaughter of German civilians. An Operational Research Section which was supposed to give him independent scientific advice but was too timid to challenge any essential element of his policies. A collection of staff officers at the Command headquarters who reminded me, when occasionally I was invited to go and have a drink with them at the officers' mess, of the Oxford dons that the historian Edward Gibbon described two hundred years ago in his autobiography: "Their dull and deep potations excused the brisk intemperance of youth."

Many of these evils existed in military establishments long before warfare became technological. Our commander in chief was a typical example of a prescientific military man. He was brutal and unimaginative, but at least he was human and he was willing to take responsibility for the evil that he did. In himself he was not worse than General Sherman, who also did evil in a just cause. He was only carrying out, with greater enthusiasm than the situation demanded, the policy laid down by his government. His personality was not the root of the evil at Bomber Command.

The root of the evil was the doctrine of strategic bombing, which had guided the evolution of Bomber Command from its beginning in 1936. The doctrine of strategic bombing declared that the only way to win wars or to prevent wars was to rain down death and destruction upon enemy countries from the sky. This doctrine was attractive to political and military leaders in the 1930s, for two reasons. First, it promised them escape from their worst nightmare, a repetition of the frightful trench warfare of the First World War through which they had all lived. Second, it offered them a hope that war could be avoided altogether by the operation of the principle that later came to be known as "deterrence." The doctrine held that all governments would be deterred from starting wars if they knew that the consequence would be certain and ruinous bombardment. So far as the war against Germany was concerned, history proved the theory wrong on both counts. Strategic bombing neither deterred the war nor won it. There has never yet been a war that strategic bombing by itself has won. In spite of the clear evidence of history, the strategic bombing doctrine flourished in Bomber Command throughout the Second World War. And it flourishes still, in bigger countries, with bigger bombs.

Bomber Command was an early example of the new evil that science and technology have added to the old evils of soldiering. Technology has made evil anonymous. Through science and technology, evil is organized bureaucratically so that no individual is responsible for what happens. Neither the boy in the Lancaster aiming his bombs at an ill-defined splodge on his radar screen, nor the operations officer shuffling papers at squadron headquarters, nor I sitting in my little office in the Operational Research Section and calculating probabilities, had any feeling of personal responsibility. None of us ever saw the people we killed. None of us particularly cared.

The last spring of the war was the most desolate. Even after Dresden, through March and April of 1945, the bombing of cities continued. The German night fighters fought to the end, and still shot down hundreds of Lancasters in those final weeks. I began to look backward and to ask myself how it happened that I let myself become involved in this crazy game of murder. Since the beginning of the war I had been retreating step by step from one moral position to another, until at the end I had no moral position at all. At the beginning of the war I believed fiercely in the brotherhood of man, called myself a follower of Gandhi, and was morally opposed to all violence. After a year of war I retreated and said, Unfortunately nonviolent resistance against Hitler is impracticable, but I am still morally opposed to bombing. A few years later I said, Unfortunately it seems that bombing is necessary in order to win the war, and so I am willing to go to work for Bomber Command, but I am still morally opposed to bombing cities indiscriminately. After I arrived at Bomber Command I said, Unfortunately it turns out that we are after all bombing cities indiscriminately, but this is morally justified as it is helping to win the war. A year later I said, Unfortunately it seems that our bombing is not really helping to win the war, but at least I am morally justified in working to save the lives of the bomber crews. In the last spring of the war I could no longer find any excuses. Mike had fought single-handed the battle of the escape hatches and had indeed saved many lives. I had saved none. I had surrendered one moral principle after another, and in the end it was all for nothing. In that last spring, I watched the woods come to life outside the window of my office at the Command headquarters. I had a volume of the poet Hopkins on my desk. His last desperate sonnets spoke to my despair.

> See, banks and brakes
> Now, leaved how thick! Laced they are again
> With fretty chervil, look, and fresh wind shakes
> Them; birds build—but not I build; no, but strain,
> Time's eunuch, and not breed one work that wakes.
> Mine, O Thou lord of life, send my roots rain.

Thirty years later I stood with my wife and children in the air raid shelter in the garden of my wife's uncle's home in East Germany. My wife's uncle had built the shelter solidly out of brick and steel. Several bomb craters could still be traced in the ground nearby. After thirty years the roof of the shelter was still sound and the floor dry. The house stands in a village southwest of Berlin. During the years I was at Bomber Command, my wife lived in that house. She was still a child. The nights when the bombers came over she spent in the shelter. No doubt she was sitting there the night Wing Commander

MacGown came over, when I was drinking beer with the boys at Wyton. We tried without much success to explain all this to the children. "You mean Mummy was sitting down here because Daddy's friends were dropping the bombs on the garden?" You really cannot explain things like that to a seven-year-old.

4

The Blood of a Poet

During the time I was at Bomber Command, one of the London theaters put on John Drinkwater's play *Abraham Lincoln*. Drinkwater wrote it in 1918, when England was in the throes of another war. It is a thoughtful play, using the character of Lincoln to illuminate questions which were tormenting Londoners in 1918 and again in 1944. Is there such a thing as a just war? Does any cause, no matter how just, justify the tragedy and barbarity that war brings with it? In those bleak times, Londoners were hungry for answers to such questions, and the play did well at the box office. The fact that the hero was an American may have helped. We were not in a mood to accept any of our own politicians as heroes. Lincoln was like Gandhi, remote enough to be credible.

We had not been overexposed to American history in our school days, and so we responded naïvely and intensely to scenes that would make a native American yawn. The high point of the drama comes in the last scene but one, at the courthouse in Appomattox, when the immaculate Lee walks in to surrender to the disheveled Grant. After Lee departs, Grant relaxes with Meade and they discuss the reasons why they finally won the war. "We've had courage and determination," says Grant. "And we've had wits, to beat a great soldier. I'd say that to any man. But it's Abraham Lincoln, Meade, who has kept us a great cause clean to fight for. It does a man's heart good to know he's given victory to such a man to handle. A glass, Meade? [Pouring out whisky]." Whether Grant in real life ever said these words to Meade I had no means of knowing. Nor did it matter. What mattered was that in 1865, at the end of a long and bitter war, somebody might have used these words without hypocrisy. A great cause clean to fight for. Lincoln had understood that it was important, not just to win his war, but to win it so far as possible with clean hands. Our leaders in 1944 had no such understanding. In 1944 we were well set to win our war, which we had begun in 1939 with a good enough cause. But we were also well set on the path which led to Dresden, to Hiroshima, to the nuclear terror in which the whole world now lives. We had dirtied a good cause, and the dirt stuck to us. It was just as Edith Nesbit said when she wrote the rules of the magic city. We had wished for a force of strategic bombers to fight our war for us, and so we were condemned to live with strategic bomber forces for the rest of our lives.

A few days after the destruction of Dresden, our daily newspaper, the *News-Chronicle*, reported the death of Frank Thompson. This was no ordinary death. But to explain the meaning of this death I must go back again to 1936, when I was twelve and Frank was fifteen.

One of the virtues of the school at Winchester where Frank and I were boarders was that boys of all ages were thrown together in big rooms, ten or twenty to a room. There was no privacy for anybody. The buildings were 550 years old and we lived in them as our four-teenth-century predecessors had lived, in a constant and cheerful uproar. Coming into this bedlam as a shrimpy twelve-year-old with a treble voice, I crept into a corner, wondered and watched and listened. My main concern was to avoid being stepped on in the verbal and physical battles that unpredictably raged around the room. It was like that marvelous Russian film *The Childhood of Maxim Gorky*, made in 1938 with Mark Donskoy as director. Alyosha Lyarsky plays the child Gorky, trying to survive in a small house crowded with a family of quarreling Russian peasants. Whenever I get a chance to see that film it reminds me of Frank and of my early days at Winchester. Among the boys in our room, Frank was the largest, the loudest, the most uninhibited and the most brilliant. So it happened that I came to know Frank very well, and learned from him more than I learned from anybody else at that school, although he may scarcely have been aware of my existence. One of my most vivid memories is of Frank coming back from a weekend in Oxford, striding into our room and singing at the top of his voice, "She's got . . . what it takes." This set him apart from the majority in our cloistered all-male society.

At fifteen, Frank had already won the title of College Poet. He was a connoisseur of Latin and Greek literature and could talk for hours about the fine points of an ode of Horace or of Pindar. Unlike the other classical scholars in our crowd, he also read medieval Latin and modern Greek. These were for him not dead but living languages. He was more deeply concerned than the rest of us with the big world outside, with the civil war then raging in Spain, with the world war that he saw coming. From him I caught my first inkling of the great moral questions of war and peace that were to dominate our lives ever afterward. Listening to him talking, I learned that there is no way to rightly grasp these great questions except through poetry. For him, poetry was no mere intellectual amusement. Poetry was man's best effort down the ages to distill some wisdom from the inarticulate depths of his soul. Frank could no more live without poetry than I could live without mathematics.

Frank wrote little before he died, and published less. I quote here only one of his poems, addressed directly to the theme of war. It was written in 1940, shortly after the British Army was driven out of France. Frank sees this event through the eyes of the Chorus in the *Agamemnon* of Aeschylus. The chorus of citizens of Argos is brooding upon the ten-year war as they wait for the return of Agamemnon to his home after the fall of Troy. To Frank it is obvious and natural that the grief and hatred of these Greeks of three thousand years ago, made immortal by a great poet six hundred years later, should mirror and illuminate our own anguish. The essentials of war, the human passion and tragedy, are the same, whether it is the war of Troy or the war of Dunkirk. So Frank weaves these two wars together in his poem, using lines from the Aeschylus Chorus at the ends of his stanzas. The poem is called "Allotrias diai Gynaikos (For the sake of another man's wife), *Agamemnon* 437–451."

> Between the dartboard and the empty fireplace
> They are talking of the boys the village has lost;
> Tom, our best bowler all last season,
> Died clean and swift when his plane went reeling;
> Bill, who drank beer and laughed, is now asleep
> Behind Dunkirk, helped others to escape;
> And Dave went down on that aircraft carrier,
> Dave, whom nobody minded,
> But who played the flute rather well, I remember.

"These boys died bravely. We'll always be proud of them,
They've given old Adolf something to set him thinking."
That was the loudest, the driving wave of opinion.
But in the corners hear the eddies singing—
"For the sake of another man's wife."
They died in a war of others' making.

"Helen the Fair went over the water
With Paris your friend, one of your own gang,
Whom we never trusted, but you feasted
For years with fawning, let your lands go hang.
We warned you. You could have stopped it. . . .
But now we have sent *our* sons from the cornfields.
War, like a grocer, weighs and sends us back
Ashes for men, and all our year goes black.

"Yes. They died well, but not to suit your purpose;
Not so that you could go hunting with two horses,
While their sons touched their caps, opening gates for pennies.
Perhaps we shall take a hand, write our own ending."
One growls this beneath his breath.
Soft, but the Titan heard it waking.

Frank was sensitive enough to feel the enchantment of Winchester but strong enough to react against it. "The culture one imbibed at Winchester," he wrote later, "was too nostalgic. Amid those old buildings and under those graceful lime-trees it was easy to give one's heart to the Middle Ages and believe that the world had lost its manhood along with Abelard. One fell in love with the beauty of the past, and there was no dialectician there to explain that the chief glory of the past was its triumph over the age that came before it, that Abelard was great because he was a revolutionary." Frank, at any rate, did not content himself with studying the past. He persuaded one of the teachers to give regular classes in Russian and quickly became fluent in it, finding the modern revolutionary poets, Gusyev and Mayakovsky, more to his taste than the classics. I later joined the class and so was able to share at least this one of Frank's enthusiasms. But his appetite for languages was insatiable. He started an "Obscure Languages Club" among the boys in our room, who then competed with one another in trading insults and obscenities in as many different languages as possible. For a while there was a project to write Russian verses in Glagolitic script, a wonderfully ornate and curlicued alphabet that flourished briefly in the Dark Ages before the practical Saint Cyril replaced it with Cyrillic. "All the Slav languages are good," Frank wrote, "but beside Russian, Polish and Czech seem nervous and restless, Bulgarian poor and untutored, and Serbo-Croat, which is probably the next most satisfactory, just a little barbarous—a fine language for guerrillas and men who drink slivovitz in the mountains, not yet fitted for the complex philosophies of our times. But Russian is a sad, powerful language and flows gently off the tongue like molten gold." Later he changed his mind about Bulgarian.

I saw no more of Frank after he left Winchester in 1938. He went to Oxford, joined the Communist Party, enlisted in the army when war began in 1939, and spent most of the war years in the Middle East. He was in Libya, Egypt, Palestine and Persia, occasionally fighting and always adding to his stock of friends and languages. In January 1944 he was dropped by parachute into German-occupied Yugoslavia. His mission was to make contact and serve as British liaison officer with the underground resistance movement in Bulgaria, organizing air drop support and radio communications with the Allied Command in Cairo. In his last letter home, in April, he wrote, "There isn't really any news about myself. I've been working hard, I hope to some purpose, and keeping brave company, some of the best in the world. Next to this comradeship, my greatest pleasure has been rediscovering things like violets, cowslips and plum-blossom after three lost springs."

We read the end of the story in the *News-Chronicle* almost a year later. One of the Bulgarian delegates to the World Trade Union Congress in London had been an eyewitness.

Major Frank Thompson was executed about June 10 after a mock trial at Litakovo. He had been in captivity about ten days. With him perished four other officers, one American, a Serb and two Bulgarians, and eight other prisoners.

A public "trial" was hastily staged in the village hall. The hall was packed with spectators. The eyewitness saw Frank Thompson sitting against a pillar smoking his pipe. When he was called for questioning, to everyone's astonishment he needed no interpreter but spoke in correct and idiomatic Bulgarian. "By what right do you, an Englishman, enter our country and wage war against us?" he was asked. Major Thompson answered, "I came because this war is something very much deeper than a struggle of nation against nation. The greatest thing in the world now is the struggle of Anti-Fascism against Fascism." "Do you not know that we shoot men who hold your opinion?" "I am ready to die for freedom. And I am proud to die with Bulgarian patriots as companions." . . .

Major Thompson then took charge of the condemned and led them to the castle. As they marched off before the assembled people he raised the salute of the Fatherland Front which the Allies were helping, the clenched fist. A gendarme struck his hand down. But Thompson called out to the people, "I give you the salute of freedom." All the men died raising this salute. The spectators were sobbing. Many present declared that the scene was one of the most moving in all Bulgarian history, and that the men's amazing courage was the work of the English officer who carried their spirits as well as his own.

Everything in this account rings true except for one word. The word "Anti-Fascism" is, I suspect, a euphemism supplied by the Bulgarian trade union delegate. Frank always called a spade a spade. I am almost sure that he really said, "The greatest thing in the world now is the struggle of Communism against Fascism." He was, after all, a Communist. His Bulgarian comrades were Communists. They did not live long enough to discover that communism and freedom are not always synonymous. Communism was for Frank not the communism of the intellectuals but the communism of the Soviet truck-driver whom he met once by chance taking a convoy of trucks through a mountain valley in Persia. Here is Frank's account of their meeting.

"H'are you doing?" I shouted at him in Russian. His grin broadened as he heard his own tongue. He came slowly towards us. "How am I doing? Well. Very well." He came and leaned on the door of my truck, grinning thoughtfully, feeling none of our Western obligation to continue conversation. "Splendid news from Kavkaz," I said. We had just heard of the first victories at Ordzhonokidze. He grinned again. "You think it is good?" "Yes. Very good. Don't you?" He thought and grinned and looked steadily at me for nearly half a minute. "Yes, it is very good." Another half-minute devoted to thinking and grinning. "Yes, it is just as Comrade Stalin said. He said, 'There'll be a holiday on our street, too.' And so there will! So there will! There'll be a holiday on our street, too!" We both laughed at this. "Yes!" I said. "So there will! There'll be a holiday on our street, too!" The traffic cleared and we moved on. But for hours after, my inner heart laughed and sang as it hadn't done for months.

The same laughing and singing must have been in Frank's heart as he gave the clenched fist salute to the crowd in Litakovo. In September 1944, Soviet troops entered Bulgaria, the Fatherland Front took over the government, and Frank was proclaimed a national hero. The railway station of Prokopnik, where the partisans had fought one of their fiercest battles, was renamed Major Thompson Station. He lies now with his comrades on a hilltop above Litakovo village, under a stone with an inscription from the Bulgarian poet Christo Botev:

> I may die very young
> But I shall be satisfied
> If my people later say
> "He died for justice,
> For justice and for liberty."

The news of Frank's death came too late to make any change in the routine of my life at Bomber Command. I continued, during the final months of the war in Europe, to do what I could as a technician to bring the bombers safely home from their missions. But it became clearer and clearer as the weeks went by that our bombing of cities was a pointless waste of lives. Four weeks after Dresden we attacked the ancient cathedral city of Würzburg and shattered one of the finest Tiepolo ceilings of Europe in the bishop's palace. The bomber crews were particularly happy to obliterate Würzburg because they knew that the deadly German tracking and fire-control radars were called Würzburg radars. Nobody told the crews that the city of Würzburg had as much to do with the radars as our own cathedral city of Winchester had to do with Winchester rifles. I began more and more to envy the technicians on the other side who were helping the German night fighter crews to defend their homes and families. The night fighters and their supporting organization put up an astonishing performance, continuing to fight and to cause us serious losses until their last airfields were overrun and Hitler's Germany ceased to exist. They ended the war morally undefeated. They had the advantage of knowing what they were fighting for. Not, in those last weeks of the war, for Hitler, but for the preservation of what was left of their cities and their people. We had given them at the end of the war the one thing that they lacked at the beginning, a cause clean to fight for.

I also envied Frank. Not that I altogether believed in the cause he died for. In 1945 I could already see that the government he helped install in Bulgaria was unlikely to fulfill the hopes he had held for it. It was undoubtedly better in many respects than the government it overthrew. But it was not, and could not have been, a government of justice and liberty. In 1943 Frank had written, "There is a spirit abroad in Europe which is finer and braver than anything that tired continent has known for centuries, and which cannot be withstood. It is the confident will of whole peoples, who have known the utmost humiliation and suffering and have triumphed over it, to build their own life once and for all." It may be difficult, thirty years later, to find much evidence for this spirit in the bureaucrats who are now running the government in Sofia. But I have no doubt whatever that this spirit existed among the Bulgarian partisans with whom Frank lived and fought. And the mere fact that they fought and died in this spirit gave a lasting historical legitimacy to the government they established. However imperfect that government may be as an embodiment of their ideals, the monument on the hill at Litakovo remains as a challenge to future generations to prove that they did not die in vain.

It is a common irony of history that the great prophets often misjudge the place of their ultimate triumph. The Buddha failed to hold India and is revered in Japan. Marx failed to make a revolution in Germany and succeeded against all his expectation in Russia. Likewise, Frank's dream, "the confident will of whole peoples, who have known the utmost humiliation and suffering and have triumphed over it, to build their own life once and for all," has failed to come to fruition as he expected in Europe, but it has been magnificently successful as the driving force of political change almost everywhere else—in China, in Africa, in Vietnam, and among the black people of America. I was not wise enough to foresee all these events in the spring of 1945, but I knew already then that Frank had died for a dream larger than Bulgarian politics. I knew that if any hope of salvation for mankind was to emerge from the wreckage of World War II, that hope could come only from the poet's war that Frank fought, not from the technician's war that I was engaged in. It was easy, at that moment in history, to envy the dead.

What lasting lesson can we learn from these experiences? For me, at least the main lesson is clear. A good cause can become bad if we fight for it with means that are indiscriminately murderous. A bad cause can become good if enough people fight for it in a spirit of comradeship and self-sacrifice. In the end it is how you fight, as much as why you fight, that makes your cause good or bad. And the more technological the war becomes, the more disastrously a bad choice of means will change a good cause into evil. I learned this lesson from my years at Bomber Command, and from the example of Frank's life and death. Unfortunately, many of my generation who were on the winning side in World War II did not learn this lesson. If they had learned it, they would not have led us to disaster twenty years later in Vietnam. I had the advantage, when the American bombers began bombing in Vietnam, of knowing that our cause was hopeless, because I knew that Frank's spirit was out there in the jungle fighting for Ho Chi Minh.

The Americans in 1945 went through an experience directly opposite to mine. I had taken part in a bombing campaign which caused us enormous losses and failed to achieve any decisive result. I came to the end of it aware that the German defenses had by and large defeated us. The Americans began their campaign of indiscriminate bombing of cities in Japan just as we were finishing ours in Germany. Their Twenty-first Bomber Command, commanded by General Curtis LeMay and based in the Mariana Islands, attacked Tokyo with fire bombs three weeks after we attacked Dresden and achieved equally spectacular results. In this, the first raid of their campaign, they raised in Tokyo the fire storm that we never achieved in Berlin. They killed 130,000 people and destroyed half of the city in one night, losing only fourteen planes. They continued the campaign in the same style for three months and paused on June 15 because they had run out of cities to burn. The defenses were ineffective and the bomber losses were militarily inconsequential. The urban economy of Japan was shattered. Whether the Japanese industrial machine would have recovered, given time, as the German industries recovered from repeated bombings, we shall never know. The Japanese were not given time. The American bombing campaign was as clear a victory as ours was a clear defeat. Unfortunately, people learn from defeat more than they learn from victory.

While the fire bombing of Japan was in progress, the scientists at Los Alamos were putting the finishing touches to their first atomic bombs, and Secretary of War Stimson was meeting with his advisers to decide how these bombs were to be used. At the time I knew nothing of their activities. All that I knew about such matters was contained in a book which I had ordered from my Cambridge University Press catalog in the old prewar days in Winchester, a book called *New Pathways in Science,* by the astronomer Arthur Eddington. Eddington's book, published in 1935, has a chapter on "Subatomic Energy." In this chapter are two sentences which had impressed me deeply. Through the long years of war I had kept them in mind.

I have referred to the practical utilization of subatomic energy as an illusive hope which it would be wrong to encourage; but in the present state of the world it is rather a threat which it would be a grave responsibility to disparage altogether. It cannot be denied that for a society which has to create scarcity to save its members from starvation, to whom abundance spells disaster, and to whom unlimited energy means unlimited power for war and destruction, there is an ominous cloud in the distance though at present it be no bigger than a man's hand.

Henry Stimson and his advisers were not insensitive to the moral issues with which they were confronted. The record of their deliberations leaves no doubt that they agonized long and hard over the decision to use the bombs, and that they recognized the historic importance of their decision. They had to balance the overwhelming short-term value of a quick and decisive end to the war against the long-term and uncertain dangers to mankind that would follow from the establishment of a precedent for actual use of nuclear weapons. It is still possible to argue that they made the right decision. Many books have been written analyzing their decision with the wisdom of hindsight, in the light of knowledge which they did not possess concerning the political forces that were then struggling within the Japanese government. Nobody doubts that the decision was made by men who sincerely believed that it would save many thousands of lives, Japanese as well as American, that would otherwise have been sacrificed in the continuation of the war.

Two factors made it almost inevitable that Stimson, and President Truman following Stimson's advice, would decide to use the bombs. First was the fact that the whole apparatus for delivering the bombs —the B-29 bombers, the strategic bomber bases in the Mariana Islands, the trained crews, and the bureaucratic machinery of the Twenty-first Bomber Command—already existed. The B-29 had been designed and built for the specific purpose of bombing Japan from distant island bases. Not to use all this apparatus, when it was there ready and waiting for the word Go, would have been a hard decision to make and harder still to justify to the American public if the war had continued. The second factor prejudicing Stimson's decision was the fact that indiscriminate fire bombing of Japanese cities had already occurred and was widely approved. Stimson was well aware of the enormous quantitative difference in destructive potential between nuclear and conventional bombs, but it was difficult for him to feel that there was a difference in the quality of evil between the killing of 130,000 people by old-fashioned fire bombs in Tokyo and the killing of about the same number by a nuclear bomb in Hiroshima. Those who argued against the use of nuclear weapons could only speak about long-range consequences and dangers. They could not say simply, "We should not do this because it is wrong," unless they were also prepared to put a stop to indiscriminate use of conventional bombs. The ground on which Stimson might have been able to make a moral stand was already surrendered when the fire bombing started in March. Long before that, in England and in America independently, the moral issues had been effectively prejudged when the decisions were made to build strategic bomber forces and to wage war with them against civilian populations. Hiroshima was only an afterthought.

Two weeks before he parachuted into Yugoslavia in 1944, Frank Thompson wrote from Cairo, "Yesterday I read over Lincoln's Second Inaugural, which is, I suppose, when one considers the circumstances in which it was written, one of the most remarkable speeches in human history. It made me think that, if anyone wanted to find a classic example of Divine Nemesis, what better than our present war against Japan? All our filthy record in the Far East, beginning with the Opium Wars, being paid for now in rivers of blood." "Fondly do we hope, fervently do we pray," Lincoln had said, "that this mighty scourge of war may speedily pass away. Yet, if God wills that it continue until all the wealth piled by the bondsman's two hundred and fifty years of unrequited toil shall be sunk, and until every drop of blood drawn with the lash shall be paid by another drawn with the sword—as was said three thousand years ago, so still it must be said, 'The judgments of the Lord are true and righteous altogether.' "

In August 1945 I was still working at Bomber Command. After the war in Europe ended, it was decided that a force of British bombers should be sent to bases in Okinawa, from which they would add their token contribution to the American strategic bombing of Japan. I was supposed to fly out with them to Okinawa. Then, on August 7, the *News-Chronicle* arrived at my breakfast table in London with the giant headline "New Force of Nature Harnessed." I liked that. It was big and impersonal. It was childhood's end. Now perhaps we could all start behaving like grownups. Whoever wrote that headline understood that this was something bigger than one side winning a victory in a tribal squabble. It meant, with luck, that we would be finished once and for all with strategic bombing campaigns. I agreed emphatically with Henry Stimson. Once we had got ourselves into the business of bombing cities, we might as well do the job competently and get it over with.

I felt better that morning than I had felt for years. I did not bother to go to the office. Those fellows who had built the atomic bomb obviously knew their stuff. They must be an outstandingly competent bunch of people. The thought occurred to me that I might one day get to meet them. I would enjoy that. I had spent too long messing around with stupid old-fashioned bombs. It was easy, in the happiness of that August morning, to forget what Grant said to Meade at Appomattox, to forget the *Agamemnon* Chorus, to forget the Bulgarian partisans, to forget Frank's clenched fist salute and the memorial stone at Litakovo, to forget Eddington's warning, to forget Lincoln's Second Inaugural, to forget the agony of the people still slowly dying of burns and radiation sickness in Hiroshima. Later, much later, I would remember these things.

II. AMERICA

Where does one go from a world of insanity?
Somewhere on the other side of despair.

T. S. ELIOT, The Family Reunion, 1939

5

A Scientific Apprenticeship

In September 1947 I enrolled as a graduate student in the physics department of Cornell University at Ithaca. I went there to learn how to do research in physics under the guidance of Hans Bethe. Bethe is not only a great physicist but also an outstanding trainer of students. When I arrived at Cornell and introduced myself to the great man, two things about him immediately impressed me. First, there was a lot of mud on his shoes. Second, the other students called him Hans. I had never seen anything like that in England. In England, professors were treated with respect and wore clean shoes.

Within a few days Hans found me a good problem to work on. He had an amazing ability to choose good problems, not too hard and not too easy, for students of widely varying skills and interests. He had eight or ten students doing research problems and never seemed to find it a strain to keep us busy and happy. He ate lunch with us at the cafeteria almost every day. After a few hours of conversation, he could judge accurately what each student was capable of doing. It had been arranged that I would only be at Cornell for nine months, and so he gave me a problem that he knew I could finish within that time. It worked out exactly as he said it would.

I was lucky to arrive at Cornell at that particular moment. Nineteen forty-seven was the year of the postwar flowering of physics, when new ideas and new experiments were sprouting everywhere from seeds that had lain dormant through the war. The scientists who had spent the war years at places like Bomber Command headquarters and Los Alamos came back to the universities impatient to get started again in pure science. They were in a hurry to make up for the years they had lost, and they went to work with energy and enthusiasm. Pure science in 1947 was starting to hum. And right in the middle of the renascence of pure physics was Hans Bethe.

At that time there was a single central unsolved problem that absorbed the attention of a large fraction of physicists. We called it the quantum electrodynamics problem. The problem was simply that there existed no accurate theory to describe the everyday behavior of atoms and electrons emitting and absorbing light. Quantum electrodynamics was the name of the missing theory. It was called quantum because it had to take into account the quantum nature of light, electro because it had to deal with electrons, and dynamics because it had to describe forces and motions. We had inherited from the prewar generation of physicists, Einstein and Bohr and Heisenberg and Dirac, the basic ideas for such a theory. But the basic ideas were not enough. The basic ideas could tell you roughly how an atom would behave. But we wanted to be able to calculate the behavior exactly. Of course it often happens in science that things are too complicated to be calculated exactly, so that one has to be content with a rough qualitative understanding. The strange thing in 1947 was that even the simplest and most elementary objects, hydrogen atoms and light quanta, could not be accurately understood. Hans Bethe was convinced that a correct and exact theory would emerge if we could figure out how to calculate consistently using the old prewar ideas. He stood like Moses on the mountain showing us the promised land. It was for us students to move in and make ourselves at home there.

A few months before I arrived at Cornell, two important things had happened. First, there were some experiments at Columbia University in New York which measured the behavior of an electron a thousand times more accurately than it had been measured before. This made the problem of creating an accurate theory far more urgent and gave the theorists some accurate numbers which they had to try to explain. Second, Hans Bethe himself did the first theoretical calculation that went substantially beyond what had been done before the war. He calculated the energy of an electron in an atom of hydrogen and found an answer agreeing fairly well with the Columbia measurement. This showed that he was on the right track. But his calculation was still a pastiche of old ideas held together by physical intuition. It had no firm mathematical basis. And it was not even consistent with Einstein's principle of relativity. That was how things stood in September when I joined Hans's group of students.

The problem that Hans gave me was to repeat his calculation of the electron energy with the minimum changes that were needed to make it consistent with Einstein. It was an ideal problem for somebody like me, who had a good mathematical background and little knowledge of physics. I plunged in and filled hundreds of pages with calculations, learning the physics as I went along. After a few months I had an answer, again agreeing near enough with Columbia. My calculation was still a pastiche. I had not improved on Hans's calculation in any fundamental sense. I came no closer than Hans had come to a basic understanding of the electron. But those winter months of calculation had given me skill and confidence. I had mastered the tools of my trade. I was now ready to start thinking.

As a relaxation from quantum electrodynamics, I was encouraged to spend a few hours a week in the student laboratory doing experiments. These were not real research experiments. We were just going through the motions, repeating famous old experiments, knowing beforehand what the answers ought to be. The other students grumbled at having to waste their time doing Mickey Mouse experiments. But I found the experiments fascinating. In all my time in England I had never been let loose in a laboratory. All these strange objects that I had read about, crystals and magnets and prisms and spectroscopes, were actually there and could be touched and handled. It seemed like a miracle when I measured the electric voltage produced by light of various colors falling on a metal surface and found that Einstein's law of the photoelectric effect is really true. Unfortunately I came to grief on the Millikan oil drop experiment. Millikan was a great physicist at the University of Chicago who first measured the electric charge of individual electrons. He made a mist of tiny drops of oil and watched them float around under his microscope while he pulled and pushed them with strong electric fields. The drops were so small that some of them carried a net electric charge of only one or two electrons. I had my oil drops floating nicely, and then I grabbed hold of the wrong knob to adjust the electric field. They found me stretched out on the floor, and that finished my career as an experimenter.

I never regretted my brief and almost fatal exposure to experiments. This experience brought home to me as nothing else could the truth of Einstein's remark, "One may say the eternal mystery of the world is its comprehensibility." Here was I, sitting at my desk for weeks on end, doing the most elaborate and sophisticated calculations to figure out how an electron should behave. And here was the electron on my little oil drop, knowing quite well how to behave without waiting for the result of my calculation. How could one seriously believe that the electron really cared about my calculation, one way or the other? And yet the experiments at Columbia showed that it did care. Somehow or other, all this complicated mathematics that I was scribbling established rules that the electron on the oil drop was bound to follow. We know that this is so. Why it is so, why the electron pays attention to our mathematics, is a mystery that even Einstein could not fathom.

At our daily lunches with Hans we talked endlessly about physics, about the technical details and about the deep philosophical mysteries. On the whole, Hans was more interested in details than in philosophy. When I raised philosophical questions he would often say, "You ought to go and talk to Oppy about that." Oppy was Robert Oppenheimer, then newly appointed as director of the Institute for Advanced Study at Princeton. Sometime during the winter, Hans spoke with Oppy about me and they agreed that after my year at Cornell I should go for a year to Princeton. I looked forward to working with Oppy but I was also a bit scared. Oppy was already a legendary figure. He had been the originator and leader of the bomb project at Los Alamos. Hans had worked there under him as head of the Theoretical Division. Hans had enormous respect for Oppy. But he warned me not to expect an easy life at Princeton. He said Oppy did not suffer fools gladly and was sometimes hasty in deciding who was a fool.

One of our group of students at Cornell was Rossi Lomanitz, a rugged character from Oklahoma who lived in a dilapidated farmhouse outside Ithaca and was rumored to be a Communist. Lomanitz was never at Los Alamos, but he had worked with Oppy on the bomb project in California before Los Alamos was started. Being a Communist was not such a serious crime in 1947 as it became later. Seven years later, when Oppy was declared to be a Security Risk, one of the charges against him was that he had tried to stop the army from drafting Lomanitz. Mr. Robb, the prosecuting attorney at the trial, imputed sinister motives in Oppy's concern for Lomanitz. Oppy replied to Robb, "The relations between me and my students were not that I stood at the head of a class and lectured." That remark summed up exactly what made both Hans and Oppy great teachers. In 1947 security hearings and witch hunts were far from our thoughts. Rossi Lomanitz was a student just like the rest of us. And Oppy was the great national hero whose face could be seen ornamenting the covers of *Time* and *Life* magazines.

I knew before I came to Cornell that Hans had been at Los Alamos. I had not known beforehand that I would find a large fraction of the entire Los Alamos gang, with the exception of Oppy, reassembled at Cornell. Hans had been at Cornell before the war, and when he returned he found jobs for as many as possible of the bright young people he had worked with at Los Alamos. So we had at Cornell Robert Wilson, who had been head of experimental physics at Los Alamos, Philip Morrison, who had gone to the Mariana Islands to take care of the bombs that were used at Hiroshima and Nagasaki, Dick Feynman, who had been in charge of the computing center, and many others. I was amazed to see how quickly and easily I fitted in with this bunch of weaponeers whose experience of the war had been so utterly different from my own. There was endless talk about the Los Alamos days. Through all the talk shone a glow of pride and nostalgia. For every one of these people, the Los Alamos days had been a great experience, a time of hard work and comradeship and deep happiness. I had the impression that the main reason they were happy to be at Cornell was that the Cornell physics department still retained something of the Los Alamos atmosphere. I, too, could feel the vivid presence of this atmosphere. It was youth, it was exuberance, it was informality, it was a shared ambition to do great things together in science without any personal jealousies or squabbles over credit. Hans Bethe and Dick Feynman did, many years later, receive well-earned Nobel Prizes, but nobody at Cornell was grabbing for prizes or for personal glory.

The Los Alamos people did not speak in public about the technical details of bombs. It was surprisingly easy to talk around that subject without getting onto dangerous ground. Only once I embarrassed everybody at the lunch table by remarking in all innocence, "It's lucky that Eddington proved it's impossible to make a bomb out of hydrogen." There was an awkward silence and the subject of conversation was abruptly changed. In those days the existence of any thoughts about hydrogen bombs was a deadly secret. After lunch one of the students took me aside and told me in confidence that unfortunately Eddington was wrong, that a lot of work on hydrogen bombs had been done at Los Alamos, and would I please never refer to the subject again. I was pleased that they trusted me enough to let me in on the secret. After that I felt I was really one of the gang.

Many of the Los Alamos veterans were involved in political activities aimed at educating the public about the nuclear facts of life. The main thrust of their message was that the American monopoly of nuclear weapons could not last, and that in the long run the only hope of survival would lie in a complete surrender of all nuclear activities to a strong international authority. Philip Morrison was especially eloquent in spreading this message. Oppy had been saying the same thing more quietly to his friends inside the government. But by 1948 it was clear that the chance of establishing an effective international authority on the basis of the wartime Soviet-American alliance had been missed. The nuclear arms race had begun, and the idea of international control could at best be a long-range dream.

Our lunchtime conversations with Hans were often centered on Los Alamos and on the moral questions surrounding the development and use of the bomb. Hans was troubled by these questions. But few of the other Los Alamos people were troubled. It seemed that hardly anybody had been troubled until after Hiroshima. While the work was going on, they were absorbed in scientific details and totally dedicated to the technical success of the project. They were far too busy with their work to worry about the consequences. In June 1945 Oppy had been a member of the group appointed by Henry Stimson to advise him about the use of the bombs. Oppy had supported Stimson's decision to use them as they were used. But Oppy did not at that time discuss the matter with any of his colleagues at Los Alamos. Not even with Hans. That responsibility he bore alone.

In February 1948 *Time* magazine published an interview with Oppy in which appeared his famous confession, "In some sort of crude sense, which no vulgarity, no humor, no overstatement can quite extinguish, the physicists have known sin; and this is a knowledge which they cannot lose." Most of the Los Alamos people at Cornell repudiated Oppy's remark indignantly. They felt no sense of sin. They had done a difficult and necessary job to help win the war. They felt it was unfair of Oppy to weep in public over their guilt when anybody who built any kind of lethal weapons for use in war was equally guilty. I understood the anger of the Los Alamos people, but I agreed with Oppy. The sin of the physicists at Los Alamos did not lie in their having built a lethal weapon. To have built the bomb, when their country was engaged in a desperate war against Hitler's Germany, was morally justifiable. But they did not just build the bomb. They enjoyed building it. They had the best time of their lives while building it. That, I believe, is what Oppy had in mind when he said they had sinned. And he was right.

After a few months I was able to identify the quality that I found strange and attractive in the American students. They lacked the tragic sense of life which was deeply ingrained in every European of my generation. They had never lived with tragedy and had no feeling for it. Having no sense of tragedy, they also had no sense of guilt. They seemed very young and innocent although most of them were older than I was. They had come through the war without scars. Los Alamos had been for them a great lark. It left their innocence untouched. That was why they were unable to accept Oppy's statement as expressing a truth about themselves.

For Europeans the great turning point of history was the First World War, not the Second. The first war had created that tragic mood which was a part of the air we breathed long before the second war started. Oppy had grown up immersed in European culture and had acquired the tragic sense. Hans, being a European, had it too. The younger native-born Americans, with the exception of Dick Feynman, still lived in a world without shadows. Things are very different now, thirty years later. The Vietnam war produced in American life the same fundamental change of mood that the First World War produced in Europe. The young Americans of today are closer in spirit to the Europeans than to the Americans of thirty years ago. The age of innocence is now over for all of us.

Dick Feynman was in this respect, as in almost every other respect, an exception. He was a young native American who had lived with tragedy. He had loved and married a brilliant, artistic girl who was dying of TB. They knew she was dying when they married. When Dick went to work at Los Alamos, Oppy arranged for his wife to stay at a sanitarium in Albuquerque so that they could be together as much as possible. She died there, a few weeks before the war ended.

As soon as I arrived at Cornell, I became aware of Dick as the liveliest personality in our department. In many ways he reminded me of Frank Thompson. Dick was no poet and certainly no Communist. But he was like Frank in his loud voice, his quick mind, his intense interest in all kinds of things and people, his crazy jokes, and his disrespect for authority. I had a room in a student dormitory and sometimes around two o'clock in the morning I would wake up to the sound of a strange rhythm pulsating over the silent campus. That was Dick playing his bongo drums.

Dick was also a profoundly original scientist. He refused to take anybody's word for anything. This meant that he was forced to rediscover or reinvent for himself almost the whole of physics. It took him five years of concentrated work to reinvent quantum mechanics. He said that he couldn't understand the official version of quantum mechanics that was taught in textbooks, and so he had to begin afresh from the beginning. This was a heroic enterprise. He worked harder during those years than anybody else I ever knew. At the end he had a version of quantum mechanics that he could understand. He then went on to calculate with his version of quantum mechanics how an electron should behave. He was able to reproduce the result that Hans had calculated using orthodox theories a little earlier. But Dick could go much further. He calculated with his own theory fine details of the electron's behavior that Hans's method could not touch. Dick could calculate these things far more accurately, and far more easily, than anybody else could. The calculation that I did for Hans, using the orthodox theory, took me several months of work and several hundred sheets of paper. Dick could get the same answer, calculating on a blackboard, in half an hour.

So this was the situation which I found at Cornell. Hans was using the old cookbook quantum mechanics that Dick couldn't understand. Dick was using his own private quantum mechanics that nobody else could understand. They were getting the same answers whenever they calculated the same problems. And Dick could calculate a whole lot of things that Hans couldn't. It was obvious to me that Dick's theory must be fundamentally right. I decided that my main job, after I finished the calculation for Hans, must be to understand Dick and explain his ideas in a language that the rest of the world could understand.

In the spring of 1948, Hans and Dick went to a select meeting of experts arranged by Oppy at a lodge in the Pocono Mountains to discuss the quantum electrodynamics problem. I was not invited because I was not yet an expert. The Columbia experimenters were there, and Niels Bohr, and various other important physicists. The main event of the meeting was an eight-hour talk by Julian Schwinger, a young professor at Harvard who had been a student of Oppy's. Julian, it seemed, had solved the main problem. He had a new theory of quantum electrodynamics which explained all the Columbia experiments. His theory was built on orthodox principles and was a masterpiece of mathematical technique. His calculations were extremely complicated, and few in the audience stayed with him all the way through the eight-hour exposition. But Oppy understood and approved everything. After Julian had finished, it was Dick's turn. Dick tried to tell the exhausted listeners how he could explain the same experiments much more simply using his own unorthodox methods. Nobody understood a word that Dick said. At the end Oppy made some scathing comments and that was that. Dick came home from the meeting very depressed.

During the last months of my time at Cornell I made an effort to see as much of Dick as possible. The beautiful thing about Dick was that you did not have to be afraid you were wasting his time. Most scientists when you come to talk with them are very polite and let you sit down, and only after a while you notice from their bored expressions or their fidgety fingers that they are wishing you would go away. Dick was not like that. When I came to his room and he didn't want to talk he would just shout, "Go away, I'm busy," without even turning his head. So I would go away. And next time when I came and he let me sit down, I knew he was not just being polite. We talked for many hours about his private version of physics and I began finally to get the hang of it.

The reason Dick's physics was so hard for ordinary people to grasp was that he did not use equations. The usual way theoretical physics was done since the time of Newton was to begin by writing down some equations and then to work hard calculating solutions of the equations. This was the way Hans and Oppy and Julian Schwinger did physics. Dick just wrote down the solutions out of his head without ever writing down the equations. He had a physical picture of the way things happen, and the picture gave him the solutions directly with a minimum of calculation. It was no wonder that people who had spent their lives solving equations were baffled by him. Their minds were analytical; his was pictorial. My own training, since the far-off days when I struggled with Piaggio's differential equations, had been analytical. But as I listened to Dick and stared at the strange diagrams that he drew on the blackboard, I gradually absorbed some of his pictorial imagination and began to feel at home in his version of the universe.

The essence of Dick's vision was a loosening of all constraints. In orthodox physics you say, Suppose an electron is in this state at a certain time, then you calculate what it will do next by solving a certain differential equation, and from the solution of the equation you calculate what it will be doing at some later time. Instead of this, Dick said simply, the electron does whatever it likes. The electron goes all over space and time in all possible ways. It can even go backward in time whenever it chooses. If you start with an electron in this state at a certain time and you want to see whether it will be in some other state at another time, you just add together contributions from all the possible histories of the electron that take it from this state to the other. A history of the electron is any possible path in space and time, including paths zigzagging forward and back in time. The behavior of the electron is just the result of adding together all the histories according to some simple rules that Dick worked out. And the same trick works with minor changes not only for electrons but for everything else—atoms, baseballs, elephants and so on. Only for baseballs and elephants the rules are more complicated.

This sum-over-histories way of looking at things is not really so mysterious, once you get used to it. Like other profoundly original ideas, it has become slowly absorbed into the fabric of physics, so that now after thirty years it is difficult to remember why we found it at the beginning so hard to grasp. I had the enormous luck to be there at Cornell in 1948 when the idea was newborn, and to be for a short time Dick's sounding board. I witnessed the concluding stages of the five-year-long intellectual struggle by which Dick fought his way through to his unifying vision. What I saw of Dick reminded me of what I heard Keynes say of Newton six years earlier: "His peculiar gift was the power of holding continuously in his mind a purely mental problem until he had seen straight through it. I fancy his pre-eminence is due to his muscles of intuition being the strongest and most enduring with which a man has ever been gifted."

In that spring of 1948 there was another memorable event. Hans received a small package from Japan containing the first two issues of a new physics journal, *Progress of Theoretical Physics,* published in Kyoto. The two issues were printed in English on brownish paper of poor quality. They contained a total of six short articles. The first article in issue No. 2 was called "On a Relativistically Invariant Formulation of the Quantum Theory of Wave Fields," by S. Tomonaga of Tokyo University. Underneath it was a footnote saying, "Translated from the paper . . . (1943) appeared originally in Japanese." Hans gave me the article to read. It contained, set out simply and lucidly without any mathematical elaboration, the central idea of Julian Schwinger's theory. The implications of this were astonishing. Somehow or other, amid the ruin and turmoil of the war, totally isolated from the rest of the world, Tomonaga had maintained in Japan a school of research in theoretical physics that was in some respects ahead of anything existing anywhere else at that time. He had pushed on alone and laid the foundations of the new quantum electrodynamics, five years before Schwinger and without any help from the Columbia experiments. He had not, in 1943, completed the theory and developed it as a practical tool. To Schwinger rightly belongs the credit for making the theory into a coherent mathematical structure. But Tomonaga had taken the first essential step. There he was, in the spring of 1948, sitting amid the ashes and rubble of Tokyo and sending us that pathetic little package. It came to us as a voice out of the deep.

A few weeks later, Oppy received a personal letter from Tomonaga describing the more recent work of the Japanese physicists. They had been moving ahead fast in the same direction as Schwinger. Regular communications were soon established. Oppy invited Tomonaga to visit Princeton, and a succession of Tomonaga's students later came to work with us at Princeton and at Cornell. When I met Tomonaga for the first time, a letter to my parents recorded my immediate impression of him: "He is more able than either Schwinger or Feynman to talk about ideas other than his own. And he has enough of his own too. He is an exceptionally unselfish person." On his table among the physics journals was a copy of the New Testament.

6

A Ride to Albuquerque

The term at Cornell ended in June, and Hans Bethe arranged an invitation for me to go for five weeks to a summer school at the University of Michigan in Ann Arbor. Julian Schwinger would be lecturing there and would give us a leisurely account of the theory which he had sketched in his eight-hour marathon talk at the Pocono meeting. It was a great chance for me to hear Schwinger's ideas straight from the horse's mouth. But there was a gap of two weeks between the end of term and the beginning of summer school. Dick Feynman said, "I'm driving to Albuquerque. Why don't you come along?" I looked at the map and saw that Albuquerque was not directly on the way to Ann Arbor. I said yes, I'd come along.

My stay in the United States was financed by a Commonwealth Fund Fellowship awarded by the Harkness Foundation. The foundation generously included in its stipend the funds for a summer vacation. I was expected to travel across the continent and gain a wider perspective of the United States than could be seen from a single campus. A free ride to Albuquerque would make a good beginning.

I had Dick to myself most of the time for four days. Not all the time, because Dick loved to pick up hitchhikers. I enjoyed the hitchhikers too. These were American nomads, people with restless feet, moving from one place to another carelessly and without hurry. In England we have our nomadic tribe of gypsies, but they live in a closed-off world of their own. I had never spoken to a gypsy. Dick talked with these nomads as if they were old friends. They told us their adventures and Dick told them his. As we drove farther south and west, Dick's manner of speech changed. He was adapting to the accent and idiom of the people we picked up. Phrases like "I don't know noth'n" became more frequent. The closer we came to Albuquerque, the more Dick seemed to feel at ease with his surroundings.

We crossed the Mississippi at St. Louis and came through the Ozark country into Oklahoma. The Ozarks were the loveliest part of the trip, green hills covered with flowers and woods and an occasional quiet farmhouse. Oklahoma was a different world, rich and ugly, with new towns and factories springing up everywhere and bulldozers tearing up the earth. Oklahoma was in the middle of an oil boom. We were about halfway to Oklahoma City when we ran into a rainstorm. In that country, it seemed, not only the people were rough and raw, but nature too. It was my first taste of tropical rain. It made the heaviest rain I had ever seen in England look like a drizzle. We crawled for a while through the downpour and then ran into a traffic jam. Some boys told us there were six feet of water over the highway ahead of us and no way through. They said it had been raining like this for about a week. We turned around and retreated to a place called Vinita. There was nothing to do but get a room and wait for the floods to subside. The hotels were filled to capacity with stranded travelers. We were lucky to find a room, which Dick and I could share for fifty cents each. On the door was a notice that said, "This hotel is under new management, so if you're drunk you came to the wrong place." In that little room, with the rain drumming on the dirty window panes, we talked the night through. Dick talked of his dead wife, of the joy he had had in nursing her and making her last days tolerable, of the tricks they had played together on the Los Alamos security people, of her jokes and her courage. He talked of death with an easy familiarity which can come only to one who has lived with spirit unbroken through the worst that death can do. Ingmar Bergman in his film *The Seventh Seal* created the character of the juggler Jof, always joking and playing the fool, seeing visions and dreams that nobody else believes in, surviving at the end when death carries the rest away. Dick and Jof have a great deal in common. Many people at Cornell had told me Dick was crazy. In fact he was the sanest of the whole crowd.

Dick talked a great deal, that night in Vinita, about his work at Los Alamos. It was Bob Wilson, our good friend and the chief experimental physicist at Cornell, who had invited Dick to join the work on the bomb. Dick had answered at once by instinct, "No, I won't do it." Then he thought it over, and persuaded himself intellectually that he ought to work on it to make sure that Hitler did not get it first. So he joined the project, first in Princeton and then in Los Alamos. He threw himself furiously into the work and quickly became a leader. He was only twenty-six when they made him head of the computing section. The computers in those days were not electronic machines but human beings. Dick knew how to coach his team of computers so that they put their hearts and souls into the work. After he took over the section the output of computed problems went up ninefold. The section was going full steam ahead, racing against time to have all the calculations done before the first bomb test in July 1945. Dick was organizing them and cheering them on. It was like a grand boat race. They were racing so hard that nobody noticed when the Germans dropped out of the war and left them racing alone. When they passed the finish line, the day of the Trinity bomb test, Dick sat on the hood of a jeep and banged his bongo drums in joy. Only later he had time to think and to wonder whether perhaps his first instinctive answer to Bob Wilson had not been the right one. Since those days, he refused ever again to have anything to do with military work. He knew that he was too good at it and enjoyed it too much.

Dick had his own view of the future of nuclear weapons. Two illusions were current at that time. The conservative illusion was that American leadership in development and production of these weapons could be maintained indefinitely and would give America lasting military and political supremacy. The liberal illusion was that when all governments became aware of the dangers of nuclear annihilation they would abandon war as an instrument of national policy. Either way, nuclear weapons would become in some sense a guarantee of perpetual peace. Dick believed in neither illusion. He thought that wars would continue to occur from time to time, and that nuclear weapons would be used. He felt we were fools to think that we deserved to get away scot-free after letting these weapons loose in the world. He expected that somebody would sooner or later come back to give us a taste of our own medicine. He saw no reason to believe that other countries would be wiser or kinder than we had been. He found it amazing that people would go on living calmly in places like New York as if Hiroshima had never happened. As we drove through Cleveland and St. Louis, he was measuring in his mind's eye distances from ground zero, ranges of lethal radiation and blast and fire damage. His view of the future was bleak indeed. I felt as if I were taking a ride with Lot through Sodom and Gomorrah.

And yet Dick was never gloomy. He had absolute confidence in the ability of ordinary people to survive the crimes and follies of their rulers. Like Jof the juggler, he would sit quietly sharing his fresh milk and wild strawberries with his guests on the eve of the Day of Judgment. He knew how tough ordinary people are, how death and destruction often brings out the best in us.

It happened that just a year earlier, in the summer of 1947, I had lived for three weeks in a city of rubble, the bombed-out German city of Münster. The University of Münster had invited a group of foreign students to come there to give the German students their first contact with the world outside. We had a street plan of the city to help us find our way around the mountains of rubble. "Even a city of rubble," it said on the street plan, "in a time of poverty and misery, can express in the appearance of its streets and sidewalks and parks and gardens the pride and the resilience and the public spirit of its people." That was true. Every evening when the weather was not too bad, the hungry people of Münster emerged from their cellars with violins and cellos and bassoons to give first-rate orchestral concerts in the open air. One night they even put on an opera, *Cavalleria Rusticana*. The opera was not the greatest, but the theater, a grassy amphitheater overshadowed by magnificent beech and chestnut trees, and the beauty of the evening, and the silhouette of the ruined castle, amply made up for the imperfections of the performance. By that time I had become so accustomed to being hungry and walking over piles of rubble that I did not notice it any more. Even in three weeks you get completely used to living in a world of hunger and rubble. I talked to Dick about these experiences in Germany, and he said it was just as he would have expected. He could not imagine that any bombs, even nuclear bombs, could crush the spirit of humanity for long. "When you just think of all the crazy things we have survived," he said, "the atomic bomb is not such a big deal." Death is not such a big deal if you are Jof the juggler and can see the black wings of the angel of death flying over your head as you drive your wagon through the storm.

After the bombs, we talked of science. Dick and I were always disagreeing about science. We fought against each other's ideas, and that helped us both to think straight. Dick distrusted my mathematics and I distrusted his intuition. He had this wonderful vision of the world as a woven texture of world lines in space and time, with everything moving freely, and the various possible histories all added together at the end to describe what happened. It was essential to his view of things that it must be universal. It must describe every-·thing that happens in nature. You could not imagine the sum-over-histories picture being true for a part of nature and untrue for another part. You could not imagine it being true for electrons and untrue for gravity. It was a unifying principle that would either explain everything or explain nothing. And this made me profoundly skeptical. I knew how many great scientists had chased this will-o'-the-wisp of a unified theory. The ground of science was littered with the corpses of dead unified theories. Even Einstein had spent twenty years searching for a unified theory and had found nothing that satisfied him. I admired Dick tremendously, but I did not believe he could beat Einstein at his own game. Dick fought back against my skepticism, arguing that Einstein had failed because he stopped thinking in concrete physical images and became a manipulator of equations. I had to admit that was true. The great discoveries of Einstein's earlier years were all based on direct physical intuition. Einstein's later unified theories failed because they were only sets of equations without physical meaning. Dick's sum-over-histories theory was in the spirit of the young Einstein, not of the old Einstein. It was solidly rooted in physical reality. But I still argued against Dick, telling him that his theory was a magnificent dream rather than a scientific theory. Nobody but Dick could use his theory, because he was always invoking his intuition to make up the rules of the game as he went along. Until the rules were codified and made mathematically precise, I could not call it a theory.

I accepted the orthodox view of the nature of physical theories. According to the orthodox view, grand unifying principles are not theories. We may hope one day to find a grand unifying principle for the whole of physics, but that is a job for future generations. For the present, nature divides itself conveniently into well-separated domains, and we are content to understand one domain at a time. A theory is a detailed and precise description of nature that is valid in one particular domain. Theories that belong to different domains use different concepts and illuminate our world from different angles.

At present we see the world of physics divided into three principal domains. The first is the domain of the very large, massive objects, planets and stars and galaxies and the universe considered as a whole. In this domain, gravitation is the dominant force and Einstein's general relativity is the triumphantly successful theory. The second is the domain of the very small, the short-lived particles that are seen in high-energy collisions and inside the nuclei of atoms. In this domain, the strong nuclear forces are dominant and there is not yet any complete theory. Fragments of theories come and go, describing more or less satisfactorily some of the things the experimenters observe, but the domain of the very small remains today what it was in 1948, a world of its own still waiting to be thoroughly explored. Between the very large and the very small there is the third domain, the middle ground of physics. The middle ground is an enormous domain, including everything intermediate in size between an atomic nucleus and a planet. It is the domain of everyday human experience. It includes atoms and electricity, light and sound, gases, liquids and solids, chairs, tables and people. The theory that we called quantum electrodynamics was the theory of the middle ground. Its aim was to give a complete and accurate account of all physical processes in the third domain, excluding only the very large and the very small.

So Dick and I argued through the night. Dick was trying to understand the whole of physics. I was willing to settle for a theory of the middle ground alone. He was searching for general principles that would be flexible enough so that he could adapt them to anything in the universe. I was looking for a neat set of equations that would describe accurately what happens in the middle ground. We went on arguing back and forth. Looking back on the argument now from thirty years later, it is easy to see that we were both right. It is one of the special beauties of science that points of view which seem diametrically opposed turn out later, in a broader perspective, to be both right. I was right because it turns out that nature likes to be compartmentalized. The theory of quantum electrodynamics turned out to do all that I expected of it. It predicts correctly, with enormous accuracy, the results of all the experiments that have been done in the domain of the middle ground. Dick was right because it turns out that his general rules of space-time trajectories and sum-over-histories have a far wider range of validity than quantum electrodynamics. In the domain of the very small, now known as particle physics, the rigid formalism of quantum electrodynamics turned out to be useless, while Dick's flexible rules, now known as Feynman diagrams, are the first working tool of every theorist.

That stormy night in our little room in Vinita, Dick and I were not looking thirty years ahead. I knew only that somewhere hidden in Dick's ideas was the key to a theory of quantum electrodynamics simpler and more physical than Julian Schwinger's elaborate construction. Dick knew only that he had larger aims in view than tidying up Schwinger's equations. So the argument did not come to an end, but left us each going his own way.

Before dawn we succeeded in sleeping a little, and in the morning we started again in the direction of Oklahoma City. The rain continued, more gently than the day before. We came through Sapulpa, a town bursting at the seams as a result of the oil boom, and then the road was blocked again. Trying to detour, we arrived at the water's edge and saw the road disappear into a huge lake. On our way back through Sapulpa we saw a Cherokee Indian and his wife walking groggily along the roadside in the rain. They were soaked to the skin and jumped happily into the car. They were able to guide us onto an unpaved and muddy road which kept to high ground clear of the floods. They soon got dry and cheerful in the car and stayed with us most of the day. They were trying to make their way to Shawnee, where they were working in an oil-field construction camp. Somehow they had acquired five quarts of hooch whiskey, so they walked off the job in Shawnee and took the whiskey home to their family and friends in Sapulpa for a celebration. The celebration ended when the five quarts were gone, the day before we picked the pair up. The floods forced us to detour along the high ground to the north, farther and farther away from Shawnee. When the Indians finally left us and bade us a friendly goodbye, they were much farther from Shawnee than they had been when we found them.

Our last hurdle was the crossing of the Cimarron River. The river was about half a mile wide, the water brick red and flowing furiously with large standing waves. I was expecting the bridge to be swept away every minute as we crawled across it. On the other side the skies gradually cleared and we came peacefully into Texas for our last overnight stop.

The cactuses were blooming red in the desert and Dick was beside himself with joy as we sailed into Albuquerque. The sun was shining for us and the police cars were screaming their welcome. It took Dick some time to realize that the police cars were signaling to us to stop. They told us politely that we had violated all the traffic rules in the book and that we should appear forthwith before the justice of the peace. Fortunately the J.P. was on duty and could handle the case immediately. The J.P. informed us that we should pay a fine of fifty dollars, since we had been doing seventy in a twenty-mile-an-hour zone and the fine was one dollar for every mile per hour over the limit. He said that this was the largest speeding fine he had ever imposed. We had broken the Albuquerque record. Dick then put on one of his finest performances, explaining how he had driven two thousand miles from Ithaca to Albuquerque to visit this girl that he intended to marry, and telling what a great city Albuquerque was and how happy he was to be back again after being away for three years. Soon Dick and the J P were swapping stories about the wartime days in Albuquerque, and the end of it was that we were let off with a fine of $14.50, ten dollars for speeding and $4.50 for the expenses of the court. Dick and I split the fine and we all three shook hands on it. Then we said goodbye and went our separate ways.

I took a Greyhound bus to Santa Fe and made my way by easy stages back to Ann Arbor. I soon found out that the way to enjoy long bus rides is to travel at night and rest or explore the countryside by day. People talk more and are friendlier on the night runs. On the long overnight stretch from Denver to Kansas City I fell in with a couple of teen-agers, a young sailor from San Francisco and a girl from Kansas. We talked the night away, beginning with love affairs, continuing with family histories and God, and ending with politics. It occurred to me that if I had been listening to a conversation between strangers in England, the same subjects would have come up in the opposite order. The two of them were great talkers and kept it up in fine style until the sun broke through on the horizon ahead of us. At times they made me feel very old, and at times very young.

In the five weeks in Ann Arbor I made a host of new friends. The Ann Arbor summer school was in those days, as it had been in the 1930s, the main gathering place for itinerant physicists in summertime. Julian Schwinger's lectures were a marvel of polished elegance, like a difficult violin sonata played by a virtuoso, more technique than music. Fortunately, Schwinger was friendly and approachable. I could talk with him at length, and from these conversations more than from the lectures I learned how his theory was put together. In the lectures his theory was a cut diamond, brilliant and dazzling. When I talked with him in private, I saw it in the rough, the way he saw it himself before he started the cutting and polishing. In this way I was able to grasp much better his way of thinking. The Ann Arbor physicists generously gave me a room to myself on the top floor of their building. Each afternoon I hid up there under the roof for several hours and worked through every step of Schwinger's lectures and every word of our conversations. I intended to master Schwinger's techniques as I had mastered Piaggio's differential equations ten years before. Five weeks went by quickly. I filled hundreds of pages with calculations, working through various simple problems with Schwinger's methods. At the end of the summer school, I felt that I understood Schwinger's theory as well as anybody could understand it, with the possible exception of Schwinger. That was what I had come to Ann Arbor to do.

During the Ann Arbor days another beautiful thing happened. A long letter came from Münster, from one of the girls that I had got to know in the hungry time a year earlier. We had exchanged letters intermittently during the winter. She wrote in German, but the letter ended with a quotation from Yeats:

> I would spread the clothes under your feet,
>   But I am poor, and have only my dreams.
> I have spread my dreams under your feet;
> Tread softly, because you tread on my dreams.

I wondered whether a girl to whom English is a foreign language could possibly understand how good that stanza is as poetry. I decided she probably understood. I promised myself I would tread softly.

From Ann Arbor I took another Greyhound bus all the way to San Francisco. On this trip the most memorable part was the winding descent down Echo Creek from Wyoming to the Salt Lake basin. We passed through the mountain valleys in which the Mormon pioneers had settled a hundred years before. These valleys were tended and cared for like mountain valleys in Switzerland. Nowhere else in America had I seen land so cherished. You could see at once, these people believed they had reached the promised land, and they intended to leave it beautiful for their great-grandchildren.

I stayed ten days in San Francisco and Berkeley, taking a holiday from physics. I read Joyce's *Portrait of the Artist as a Young Man* and Nehru's autobiography. I explored California a little and decided I liked Utah better. Comparing the achievements of the settlers in Utah and California, who were building their civilizations at the same time, I felt that Utah achieved greatness while California had greatness thrust upon it. There is nothing in California to equal the Mormon valleys, with each village clustering around its big temple and the mountains on each side sweeping straight up to heaven.

At the beginning of September it was time to go back East. I got onto a Greyhound bus and traveled nonstop for three days and nights as far as Chicago. This time I had nobody to talk to. The roads were too bumpy for me to read, and so I sat and looked out of the window and gradually fell into a comfortable stupor. As we were droning across Nebraska on the third day, something suddenly happened. For two weeks I had not thought about physics, and now it came bursting into my consciousness like an explosion. Feynman's pictures and Schwinger's equations began sorting themselves out in my head with a clarity they had never had before. For the first time I was able to put them all together. For an hour or two I arranged and rearranged the pieces. Then I knew that they all fitted. I had no pencil or paper, but everything was so clear I did not need to write it down. Feynman and Schwinger were just looking at the same set of ideas from two different sides. Putting their methods together, you would have a theory of quantum electrodynamics that combined the mathematical precision of Schwinger with the practical flexibility of Feynman. Finally, there would be a straightforward theory of the middle ground. It was my tremendous luck that I was the only person who had had the chance to talk at length to both Schwinger and Feynman and really understand what both of them were doing. In the hour of illumination I gave thanks to my teacher Hans Bethe, who had made it possible. During the rest of the day as we watched the sun go down over the prairie, I was mapping out in my head the shape of the paper I would write when I got to Princeton. The title of the paper would be "The Radiation Theories of Tomonaga, Schwinger and

Feynman." This way I would make sure that Tomonaga got his fair share of the glory. As we moved on into Iowa, it grew dark and I had a good long sleep.

A few days later I collected my belongings from Ithaca and went on to Princeton. I had grown so attached to Greyhound buses I was almost sorry to be at the end of the journey. But I had work to do in Princeton. On a fine September morning I walked for the first time the mile and a half from my room in Princeton to the Institute for Advanced Study. It was exactly a year since I had left England to learn physics from the Americans. And now here I was a year later, walking down the road to the institute on a fine September morning to teach the great Oppenheimer how to do physics. The whole situation seemed too absurd to be credible. I pinched myself to make sure I wasn't dreaming. But the sun still shone and the birds still sang in the trees. I had better be careful, I said to myself. Tread softly, because you tread on my dreams.

7

The Ascent of F6

Seven years and the summer is over.
Seven years since the Archbishop left us,
He who was always kind to his people.
But it would not be well if he should return. . . .
For us, the poor, there is no action,
But only to wait and to witness. . . .
O Thomas Archbishop,
O Thomas our Lord, leave us and leave us be, in our humble
        and tarnished frame of existence, leave us; do not ask us
To stand to the doom on the house, the doom on the
        Archbishop, the doom on the world.

I sat in Oppenheimer's office in the fall of 1948 with these lines from
T. S. Eliot's *Murder in the Cathedral* ringing through my head. Eight
young physicists, six men and two women, were sharing the office
while the builders hurried to finish a new building with individual
offices for each of us. I wished the builders would never finish. It was
much cozier and friendlier in the big office, with the eight of us
sitting around a wooden table, chatting and getting to know one
another. We had come from many countries to the Institute for
Advanced Study, each of us invited by Oppenheimer to work under
his supervision. We were young and unencumbered with possessions.
Our few books and papers fitted easily on the table. It was lucky for
us that Oppenheimer was away in Europe and did not need his office.
For six or seven weeks we waited uneasily for his return. As the
weeks passed, his absence seemed to loom larger and larger, as the
Archbishop's absence looms in the first scene of Eliot's play, leading

up to his dramatic entry and the swiftly ensuing tragedy. We did not know in 1948 what kind of tragedy was to be played, but the feeling of impending doom was in the air.

Nineteen forty-eight was the year of disillusion for those who had hoped against hope that a lasting peace would emerge from the chaos of the Second World War. In that fall, while we were sitting helpless around Oppenheimer's table, Jews and Arabs were fighting in Palestine, Berlin was blockaded by Soviet troops and precariously supplied with the necessities of life by airlift, and the United Nations were failing to agree upon a plan for effective international control of nuclear weapons. Europe and half Asia were still in ruins, and already mankind seemed to be rushing into even vaster and more destructive follies. We made grim calculations of the probable course of events if things should go badly in Berlin. A rapid Soviet occupation of what was left of Western Europe, and atomic bombing of Soviet cities. A large part of the American public believed that their stockpile of atomic bombs was by itself enough to defeat the Soviet Union. I knew better. I knew that this was the same illusion which had led Napoleon in 1812 and Hitler in 1941 to disaster. In the fall of 1948 the danger seemed terribly real that the Americans would go the same way as Napoleon and Hitler, dreaming of a quick victory over the Soviet Union and awakening to find themselves in a war without end. I was seriously wondering whether I should go back to my parents in England or try to get them to join me in America before it was too late.

We sat in Oppenheimer's office and waited and worried. We knew that he bore a heavy responsibility, both for helping to bring new evil upon mankind and for trying to mitigate its consequences. We were glad that we had no share in his responsibility. We wanted only to be left in peace, to forget the war that we had survived, to escape the wars that were still to come. We were the women of Canterbury that Eliot uses as his chorus, standing on the steps of the cathedral.

> We have seen the young man mutilated,
> The torn girl trembling by the mill-stream.
> And meanwhile we have gone on living,
> Living and partly living. . . .

Building a partial shelter,
For sleeping and eating and drinking and laughter.

God gave us always some reason, some hope; but now a new
    terror has soiled us, which none can avert, none can
    avoid, flowing under our feet and over the sky.

Oddly enough, Eliot himself was also at the institute, invited like us by Oppenheimer. Prim and shy, Eliot appeared each day in the lounge at teatime, sitting by himself with a newspaper and a teacup. It was thirteen years since he had written *Murder in the Cathedral*. I wondered if he had any inkling of my private thoughts. Could this man, who had created out of the depths of his faith and despair the drama of the doomed Archbishop, be deaf to the echoes of his own words reverberating through our tragic century? Was he, too, waiting in terror and anguish for some portent of evil attending Oppenheimer's return? I never had the courage to ask him. None of our gang of young scientists succeeded in penetrating the barrier of fame and reserve that surrounded Eliot like a glass case around a mummy.

Finally, Oppenheimer returned. We were driven out of our Garden of Eden in his office and exiled to our new building. He did not say, like the Archbishop:

Peace. And let them be, in their exaltation.
They speak better than they know, and beyond your understanding.

He had no memorable words of greeting for us. Indeed, he had very little time for us at all, but rushed off almost at once to attend to some political business in Washington. His quick departure was for us a disappointment, but also a relief. We could get on with our work just as well without him. It soon became clear that I had made a mistake in trying to cast him in the role of Eliot's Archbishop. Whatever his ultimate fate might be, it would not be a traditional martyrdom. The matter was summed up well by two small boys overheard in conversation as they walked by our building. The building has a spire and a slightly ecclesiastical aura. "Is that a church?" said one small boy. "No, that's the institute," said the other. "The institute isn't a church, it's a place to eat." Oppenheimer heard of this conversation later and was delighted. He vigorously repudiated attempts of his uncritical admirers to turn him into a saint. In 1964 a German writer wrote an adulatory play for television, dramatizing Oppenheimer's trial and condemnation. Oppenheimer tried in vain to stop the production of the play by suing the producers on the grounds that it presented him in a false light. "They wanted to make that affair into a tragedy," said Oppenheimer, "but it was actually a farce."

Eliot was still at the institute when the news arrived that he had been awarded the Nobel Prize for Literature. Newspapermen swarmed around him and he retreated even further into his shell. The last time I saw him was at the grand farewell party that Oppenheimer gave in his honor before he departed on his way to Stockholm. It was a stand-up supper for about a hundred people. Oppenheimer was resplendent in black tie and tuxedo, playing to perfection the part of the gracious host. When he spoke to me, it was to give me the recipe for some excellent Mexican savories that were being served with the supper. Eliot was sequestered in a small drawing room with a group of elderly and distinguished people, apart from the main crowd. I did in the end shake Eliot's hand, but I did not find this a suitable occasion to ask him what he thought of Oppenheimer. Many years later, I asked Oppenheimer what he thought of Eliot. Oppenheimer loved Eliot's poetry and had enormous respect for his genius, but he had to admit that Eliot's stay at the institute had not been a success. "I invited Eliot here in the hope that he would produce another masterpiece, and all he did here was to work on *The Cocktail Party,* the worst thing he ever wrote."

During the anxious weeks with the young crowd in Oppenheimer's office, I had time to write the paper that set down in detail the thoughts that had come to me in the Greyhound bus in Nebraska. It was finished and sent off to the *Physical Review* before Oppenheimer returned, so that he had no chance to argue about it. After he returned, I sent him a copy of the paper and waited for something to happen. Nothing happened. That was not surprising. After all, mine was a minor contribution to the grand design of science. All I had done was to unify and tidy up the details of the quantum electrodynamics of Schwinger and Feynman. The big steps had already been taken by Schwinger and Feynman before I began. They had formulated the ideas and left to me the job of working out the equations. I knew that Oppenheimer had always been more interested in ideas than in equations. It was natural that he would have many things to do more interesting and more urgent than reading my paper.

When after some weeks I had a chance to talk to Oppenheimer, I was astonished to discover that his reasons for being uninterested in my work were quite the opposite of what I had imagined. I had expected that he would disparage my program as merely unoriginal, a minor adumbration of Schwinger and Feynman. On the contrary, he considered it to be fundamentally on the wrong track. He thought it a wasted effort to adumbrate Schwinger and Feynman, because he did not believe that the ideas of Schwinger and Feynman had much to do with reality. I had known that he never appreciated Feynman, but it came as a shock to hear him now violently opposing Schwinger, his own student, whose work he had acclaimed so enthusiastically six months earlier. He had somehow become convinced during his stay in Europe that physics was in need of radically new ideas, that this quantum electrodynamics of Schwinger and Feynman was just another misguided attempt to patch up old ideas with fancy mathematics. I was delighted to hear him talk in this style. It meant that my battle for recognition would be much more interesting. Instead of arguing with Oppenheimer about the dubious merits of my own work, I would be fighting for the entire program of quantum electrodynamics, for Schwinger's ideas and Feynman's and Tomonaga's too. Instead of fussing over details, we would be clashing on basic issues. Already I could feel that the Lord had delivered him into my hands.

Oppenheimer ran a weekly seminar, at which I took my turn as speaker. The first two occasions on which I tried to explain my ideas were disasters. After the second defeat, I reported the faltering progress of my campaign to my parents in England.

I have been observing rather carefully his behavior during seminars. If one is saying, for the benefit of the rest of the audience, things that he knows already, he cannot resist hurrying one on to something else; then when one says things that he doesn't know or immediately agree with, he breaks in before the point is fully explained with acute and sometimes devastating criticisms, to which it is impossible to reply adequately even when he is wrong. If one watches him one can see that he is moving around nervously all the time, never stops smoking, and I believe that his impatience is largely beyond his control. On Tuesday we had our fiercest public battle so far, when I criticized some unwarrantably pessimistic remarks he had made about the Schwinger theory. He came down on me like a ton of bricks, and conclusively won the argument so far as the public was concerned. However, afterwards he was very friendly and even apologized to me.

The turning point in our struggle came in the third round. My old friend and mentor Hans Bethe came down from Cornell to talk to our seminar. He wanted to speak about some calculations he had been doing with the Feynman theory. My weekly letter home describes the scene.

He was received in the style to which I am accustomed, with incessant interruptions and confused babbling of voices, and had great difficulty in making even his main points clear; while this was going on he stood very calmly and said nothing, only grinned at me as if to say "Now I see what you are up against." After that he began to make openings for me, saying in answer to a question "Well, I have no doubt Dyson will have told you all about that," at which point I was not slow to say in as deliberate a tone as possible, "I am afraid I have not got to that yet." Finally Bethe made a peroration in which he said explicitly that the Feynman theory is much the best theory and that people must learn it if they want to avoid talking nonsense; things which I have been saying for a long time but in vain.

From that point on, my path was made smooth. The next time I was scheduled to speak at the seminar, Oppenheimer actually listened. Twice more I spoke, and on the morning after my fifth talk I found in my mailbox Oppenheimer's formal note of surrender, a small piece of paper with the words "Nolo contendere. R.O." scrawled on it in his handwriting.

A few days later, Oppenheimer handed me a typed letter appointing me a long-term member of the institute and defining a generous arrangement under which I could come for periodic visits to Princeton while continuing to live in England. As he gave me the letter he delivered one of the Delphic utterances for which he was famous: "You can show this to the harbor master at Lowestoft when you start in your small boat." Perhaps he was thinking of the great physicist Niels Bohr, who escaped from German-occupied Denmark in a small boat in 1943 to get to Sweden and from there went to join Oppenheimer in Los Alamos. But why Lowestoft? I never did figure that one out.

The new year 1949 started with a mammoth meeting of the American Physical Society in New York. Oppenheimer gave a presidential address in the biggest hall, and such was the glamour of his name after his being on the cover of *Time* that the hall was packed with two thousand people half an hour before he was due to start. He spoke on the title "Fields and Quanta" and gave a very good historical summary of the vicissitudes of our attempts to understand the behavior of atoms and radiation. At the end he spoke with great enthusiasm of my work and said that it was pointing the way for the immediate future even if it did not seem deep enough to carry us farther than that. I was thinking happily to myself: Last year it was Julian Schwinger, this year it is me. Who will it be next year?

After a long winter, spring came to Princeton with a rush. Oppenheimer was spending more and more of his time in Washington. In addition to his normal government business, he was defending his friend David Lilienthal, the first chairman of the United States Atomic Energy Commission, against a vicious political attack launched by the Republicans in Congress. He defended Lilienthal skillfully and successfully. But the attack was only one of the first stirrings of the hysteria which was to lead to his own downfall five years later. While he was away in Washington, spring fever overcame our crowd of young physicists at the institute. We gave up the pretense of serious work and started to enjoy ourselves. There were many parties and expeditions to the beach. One morning scene from that spring is particularly vivid in my memory. A battered old Dodge convertible with the roof open, owned by one of the girls at the institute and driven by another, with eight or ten young institute members piled into the seats and hanging on to the back, careening at breakneck speed down through the institute woods to the river, demolishing trees and scaring to death the wild life and the distinguished professors taking their morning stroll. That scene went unrecorded in my weekly letter home. My proud parents did not need to know that I was running wild in Princeton with a bunch of young hooligans. We had never been teen-agers, having passed through that period of our lives during the years of war and deprivation, and now we were making up for the lost time. Some years later I was married to the owner of the Dodge and wrecked it on an icy road in Ithaca, but that is another story.

The end of this story is that I eventually became a professor at the institute and settled down there and lived happily ever after. I was a friend and colleague of Oppenheimer for fourteen years, from the year before his trial to the year of his death. I had plenty of time to study and reflect upon the qualities of this man who played such a paradoxical part both in my personal destiny and in the destiny of mankind. I was rarely privy to his thoughts. During the weeks of his trial, when he was staying in Washington at an address that was held tightly secret to avoid the pestering attentions of the press, my only contact with him was to deliver through a lawyer intermediary a badly needed package of laundry from Princeton. After the trial was over and the government had officially declared him untrustworthy, he came back to the institute and talked about physics. Life continued as before, except that the big steel safe and the two security guards who had watched over it night and day for seven years were no longer there. Some newspaper stories appeared, reporting rumors that the trustees of the institute were preparing to dismiss Oppenheimer on the grounds that a man publicly discredited could not adequately fulfill the duty of the director to represent the institute to the public. The trustees announced that they would be holding a meeting to review Oppenheimer's directorship and to decide upon his continued appointment. I made discreet inquiries among my friends in England to make sure I would have a job there to retreat to, so that I could resign my professorship promptly and dramatically if Oppenheimer was dismissed. The trustees held their meeting and issued a statement confirming his appointment and declaring their confidence in his leadership of the institute. I was glad to be spared the inconvenience of making a noble gesture. I was also glad that I could stay with Oppenheimer in Princeton. So far as I was concerned, he was a better director after his public humiliation than he had been before. He spent less time in Washington and more time at the institute. He was still a great public figure, a hero to the scientific brotherhood and to the international community of intellectuals, but he became more relaxed and more attentive to our day-to-day problems. He was able to get back to doing what he liked best—reading, thinking and talking about physics.

Oppenheimer had a genuine and lifelong passion for physics. He wanted always to keep struggling to understand the basic mysteries of nature. I disappointed him by not becoming a deep thinker. He had hoped, when he impulsively appointed me a long-term member of the institute, that he was securing a young Bohr or a young Einstein. If he had asked my advice at that time, I would have told him, Dick Feynman is your man, I am not. I was, and have always remained, a problem solver rather than a creator of ideas. I cannot, as Bohr and Feynman did, sit for years with my whole mind concentrated upon one deep question. I am interested in too many different things. When I came to Oppenheimer asking for guidance, he said, "Follow your own destiny." I did so, and the results did not altogether please him. I followed my destiny into pure mathematics, into nuclear engineering, into space technology and astronomy, solving problems that he rightly considered remote from the mainstream of physics. The same difference of temperament appeared in our discussions of the administration of the school of physics at the institute. He liked to concentrate new appointments in fundamental particle physics; I liked to invite people in a wide variety of specialties. So we often disagreed, but respected and understood each other better as we grew old together. We agreed on the essentials. We agreed on appointing the Chinese physicists Yang and Lee to institute professorships while they were still young, and we rejoiced together as we watched them grow over our heads into great scientific leaders.

What was so special about Oppenheimer? During the long years of daily contact I often asked myself this question. From time to time, exaggerated journalistic articles and television programs would appear, presenting him as a tragic hero. He dismissed all these effusions as unmitigated trash, but they contained a substratum of truth. I had not been altogether wrong at the beginning when I expected him to behave like the Archbishop in Eliot's play. He had a talent for self-dramatization, an ability to project to his audience an image larger than life, to bestride the world as if it were a stage. Perhaps my mistake had been only in choosing the wrong play for him to star in.

Nineteen thirty-five was a time of despair for writers all over the world. Eliot was not the only one who turned to poetic drama as the appropriate medium to express the tragic mood of that time. In the same year, *Murder in the Cathedral* appeared in England and Maxwell Anderson's *Winterset* in America. A year later, Auden and Isherwood wrote *The Ascent of F6*. *F6* was played in London in 1937 with music by Benjamin Britten and marvelously caught the shadow of coming events. *F6* is to *Murder in the Cathedral* as *Hamlet* is to *King Lear*. Eliot's Archbishop is a man of power and pride, redeemed like King Lear by serene submission to his fate in the hour of death. The hero of *F6* is a more sophisticated, more modern character. He is a mountain climber, known to his friends as M.F., a Hamlet-like figure compounded of arrogance, ambiguity and human tenderness. Over the years, as I came to know Oppenheimer better, I found many aspects of his personality foreshadowed in M.F. I came to think that *F6* was in some sense a true allegory of his life.

The plot of *F6* is simple. M.F. is an intellectual polymath, expert in European literature and Eastern philosophy. The newspaper accounts of his youthful exploits—

> Was privately educated by a Hungarian tutor.
> Climbed the west buttress of Clogwyn Du'r Arddu
> While still in his teens . . .
> Made a new traverse on the Grandes Jorasses . . .
> Studied physiology in Vienna under Niedermayer . . .
> Translated Confucius during a summer.
> Is unmarried. Hates dogs. Plays the viola da gamba.
> Is said to be an authority on Goya—

resemble strikingly the stories of Oppenheimer's precosity and preciosity as a young man. As M.F. went to the mountains for spiritual solace, so Oppenheimer went to physics. F6 is an unclimbed mountain of supreme beauty:

> Since boyhood, in dreams, I have seen the huge north face. On nights when I could not sleep I worked up those couloirs, crawled along the eastern arête, planning every movement, foreseeing every hold.

It is also a political prize important to the security of the British Empire. It stands on the frontier of the empire, adjoining the territory of a hostile power, and the natives have been led to believe that whoever first climbs the mountain shall rule over the whole region. Lord Stagmantle, representing the political establishment, offers the necessary financial support for an expedition to climb the mountain with M.F. as leader, just as General Groves offered Oppenheimer the resources of the United States Army for the project that he was to direct at Los Alamos. M.F. refuses at first to be a party to the political game, but afterward accepts the offer. As Oppenheimer said at his trial, "When you see something that is technically sweet, you go ahead and do it and you argue about what to do about it only after you have had your technical success. That is the way it was with the atomic bomb." F6 was technically sweet too.

The drama of F6 illuminates many aspects of Oppenheimer's nature: His combination of philosophical detachment with driving ambition. His dedication to pure science and his skill and self-assurance in the world of politics. His love of metaphysical poetry. His tendency to speak in cryptic poetic images. The harbor master at Lowestoft. The rapid and unpredictable shifts between warmth and coldness in his feelings toward those close to him. I once asked him if it was not a difficult thing for his children to have such a problematical figure for a father. He answered, "Oh, it's all right for them. They have no imagination." This reminded me of M.F.'s reply to the lady who accused him of being afraid when he at first refused to lead the expedition to F6:

I am afraid of a great many things, Lady Isabel, but of nothing which you in your worst nightmares could ever imagine; and of that word least of all.

At the foot of the mountain stands a monastery at which the expedition halts before beginning the ascent. The activities of the monks are directed toward the propitiation of the Demon who lives at the summit. The abbot carries a crystal ball in which each visitor looks in turn to see his personal vision of the Demon. Each sees an image of his own dreams and desires. When M.F. looks into the crystal, voices are heard coming out of the darkness offstage:

> Give me bread
> Restore my dead
> Give me a car
> Make me a star
> Make me strong
> Teach me where I belong
> Make me admired
> Make me desired
> Make us kind
> Make us of one mind
> Make us brave
> Save.

The others ask him what he sees. He says he sees nothing. Later, when he is alone with the abbot, he reveals what he saw:

Bring back the crystal. Let me look again and prove my vision a poor fake. . . . I thought I saw the raddled sick cheeks of the world light up at my approach as at the homecoming of an only son.

The abbot, whose role in the story is a little like that of Niels Bohr at Los Alamos, explains the vision:

The Demon is real. Only his ministry and his visitation are unique for every nature. To the complicated and sensitive like yourself, his disguises are more subtle. . . . I think I understand your temptation. You wish to conquer the Demon and then to save mankind.

The ascent proceeds in desperate haste as it is reported that a rival expedition is already beginning its assault on the other side of the mountain. A young climber in M.F.'s team is killed. M.F. comments:

The first victim to my pride. . . . The Abbot was perfectly right. My minor place in history is with the aberrant group of Caesars: the dullard murderers who hale the gentle from their beds of love and, with a quacking drum, escort them to the drowning ditch and the death in the desert.

So it goes on, until at the end M.F. lies dead at the summit and the monks pronounce over his body the final chorus:

> Free now from indignation,
> Immune from all frustration,
> He lies in death alone;
> Now he with secret terror
> And every minor error
> Has also made Man's weakness known.
> Whom history has deserted,
> These have their power exerted
> In one convulsive throe;
> With sudden drowning suction
> Drew him to his destruction.
> But they to dissolution go.

When I saw this play in 1937, "Whom history has deserted" meant the bankrupt political leadership of the British Empire, which was to be swept away in the approaching cataclysm of the Second World War. In 1954 the same phrase meant Lewis Strauss, the chairman of the U.S. Atomic Energy Commission, his apparatus of security officers and informers, and his allies in the press, the government and the military establishment who helped him drag Oppenheimer down in disgrace.

Auden and Isherwood succeeded remarkably in painting, or predicting, a good likeness of the character of Oppenheimer as I knew him from 1948 to 1965. But there was one essential feature missing, both from the Oppenheimer I knew and from the portrait in the play. The missing element was a greatness of spirit to which those who worked with him at Los Alamos bear almost unanimous witness. Again and again, in the reminiscences of Los Alamos veterans, we read how Oppenheimer communicated to the whole laboratory a personal style which made the enterprise run in harmony like an orchestra in the hands of a great conductor. Some of these reminiscences may be exaggerated and tinged with nostalgia. But there can be no doubt that Oppenheimer's leadership left on his Los Alamos colleagues an indelible impression of greatness. I often asked myself between 1948 and 1965: What was this greatness, and why was it no longer visible in the Oppenheimer I knew? Then at last, in 1966, I saw it for myself. In February 1966 he learned that he was dying of throat cancer. In the twelve months that remained to him, his spirit grew stronger as his bodily powers declined. The mannerisms of M.F. were discarded. He was simple, straightforward, and indomitably courageous. I saw then what his friends at Los Alamos had seen, a man carrying a crushing burden and still doing his job with such style and good humor that all of us around him felt uplifted by his example.

The last time I saw him was in February 1967, at a meeting of the faculty of the physics school at the institute. We met to decide upon the choice of visiting members for the following year. Each of us had to do a substantial amount of homework before the meeting, reading through a big brown box full of applications and judging their relative merits. Oppenheimer came to the meeting as usual, although he well knew that he would not be there to welcome the new members on their arrival. He could speak only with great difficulty, but he had done his homework and he remembered accurately the weak or strong points of the various candidates. The last words I heard him say were, "We should say yes to Weinstein. He is good." After this supreme effort of will, Robert Oppenheimer went home to his bed and collapsed into a sleep from which he never woke. He died three days later.

His wife, Kitty, called me to discuss arrangements for the memorial ceremony. Besides the music and the talks by Robert's friends describing his life and work, she wanted also to have a poem read, since poetry had always been an important part of Robert's life. She knew which poem she wanted to have read—"The Collar," by George Herbert, a poem that had been one of Robert's favorites and that she found particularly appropriate to describe how Robert had appeared to himself. Then she changed her mind. "No," she said, "that is too personal for such a public occasion." She had good reason for being afraid to bare Robert's soul in public. She knew from bitter experience how newspapers are apt to handle such disclosures. She could already imagine the horrible distortions of Robert's true feelings, appearing under the headline "Noted Scientist, Father of Atom Bomb, Turns to Religion in Last Illness." No poem was read at the ceremony.

Now Kitty is dead too, and Robert has passed beyond the reach of any further journalistic distortion. I think it will do no harm if I print Herbert's poem here in full as a memorial to both of them. Perhaps it gives us a clue to Robert's innermost nature, a hint that in his soul there was after all more of King Lear than of Hamlet.

> I struck the board, and cry'd, "No more;
> I will abroad."
> What, shall I ever sigh and pine?
> My lines and life are free; free as the road,
> Loose as the wind, and large as store.
> Shall I be still in suit?
> Have I no harvest but a thorn
> To let me bloud, and not restore
> What I have lost with cordiall fruit?
> Sure there was wine
> Before my sighs did drie it; there was corn
> Before my tears did drown it;
> Is the yeare onely lost to me?
> Have I no bayes to crown it,
> No flowers, no garlands gay? all blasted,
> All wasted?
> Not so, my heart; but there is fruit,
> And thou hast hands.
> Recover all thy sigh-blown age
> On double pleasures; leave thy cold dispute
> Of what is fit and not; forsake thy cage,
> Thy rope of sands
> Which pettie thoughts have made; and made to thee
> Good cable, to enforce and draw,
> And be thy law,
> While thou didst wink and wouldst not see.
> Away! take heed;

I will abroad.
Call in thy death's-head there, tie up thy fears;
He that forbears
To suit and serve his need
Deserves his load.
But as I rav'd and grew more fierce and wilde
At every word,
Methought I heard one calling, "Childe";
And I reply'd, "My Lord."

8

Prelude in E-Flat Minor

As a mathematically inclined child born into a musical family, I was intrigued by the intricacies of musical notations long before I developed any real understanding of music. At an early age I found my father's copy of Bach's forty-eight Preludes and Fugues for the well-tuned piano, and studied carefully the arrangements of sharps and flats in the key signatures. My father explained to me how Bach worked his way twice through all the twenty-four major and minor keys. But why is there no prelude in E-flat minor in the second book? My father did not know. Bach just decided when he came to No. 8 in the second book to write it in D-sharp minor instead. All the other key signatures come twice, but E-flat minor comes only once, at No. 8 in the first book. I was also fascinated by double sharps and double flats. Why is there a special sign for a double sharp but none for a double flat? My father did not know that either. I was giving him a hard time with my questions. I noticed that Prelude No. 3 in C-sharp major is the first one that has double sharps in it, and Prelude No. 8 in E-flat minor is the first one that has a double flat. No. 8 is special again. I asked my father to play No. 3 and No. 8 so that I could hear what double sharps and double flats sounded like. I never grew tired of hearing the delicious sound of that B double flat in Prelude No. 8.

My father was best known as a composer, but he was also in great demand as a conductor. He conducted choirs and orchestras at all levels from the local music club to the London Symphony. He accepted with good grace the fact that neither of his children inherited his musical gifts, but still he liked to take us along to listen to his concerts. At one of these concerts I was addressed by a distinguished soloist, who told me how lucky I was to be hearing so much good music at such a young age. I replied, "Music is very nice, but too long," a remark which my father gleefully repeated on many subsequent occasions. He soon discovered the way to stop me from fidgeting during the performances. He supplied me with his vocal and orchestral scores so that I could follow what was happening. I sat quiet and happy, watching in the score for the entrances of the various voices and instruments, delighting in the occasional occurrence of exotic time signatures with five or seven beats to a bar, using my eyes as a substitute for my musically defective ears.

As I grew into adolescence I began to develop a limited but genuine understanding of music. I loved to listen when my father played the piano at home for relaxation. He often played from the forty-eight Preludes and Fugues. I even learned to play some of them after a fashion myself. The Prelude in E-flat minor continued to be my favorite. Quite apart from its unique key signature and its double flat, it is also outstanding musically. It is pure Bach, and yet it carries a distinctive intensity of feeling that foreshadows Beethoven.

My father's finest hour came at the same time as England's, at the beginning of the Second World War. He was then no longer a schoolteacher. He had moved to London to be director of the Royal College of Music, one of the two major musical conservatories of England. When the war and the bombing of London began, the government and his own board of trustees urged him to evacuate the college to some safe place in the country. He refused to budge. He pointed out to his trustees that the college provided a livelihood to at least half of the leading orchestral players and concert artists of London. Most of these people came to the college to teach two or three days a week and could not live on concerts alone. If the college were evacuated, one of two consequences would follow. Either the college would lose its best teachers, or the musical life of London would be effectively closed down for the duration of the war. And in either case the careers of a whole generation of musicians would be ruined. So my father had one of the offices in the college converted into a bedroom and announced that he would stay there to keep the place running so long as any roof remained over his head. His board of trustees accepted his decision and the college stayed open. Hearing of this, the other big London conservatory, which had already made plans to evacuate, changed its mind and stayed open too. London remained musically alive, nourishing fresh talents and giving them a chance to be heard, through the six years of war. My father stayed steadfast at his post at the college, helping to put out fires on the roof at night and conducting student orchestras during the day. The only substantial loss that the college sustained was a little opera theater with an irreplaceable collection of antique operatic costumes. The damage was done at night when the professors and students were out of the building. Nobody, from the beginning of the war to the end, was injured on the premises.

During the war years I often went to lunch with my father and the professors in the college dining room. These people were hard-boiled professional musicians, averse to any display of sentiment. Their conversation consisted mainly of professional gossip and jokes. But I could feel the warmth of their loyalty to the college and the sense of comradeship that bound them and my father together. The daily experience of shared hardships and dangers created a spirit of solidarity in the college which people who have known academic institutions only in times of peace can hardly imagine. I was reminded of this spirit when I watched the citizens of bombed-out Münster perform their open-air opera in 1947, and when I heard my American friends tell tales of wartime Los Alamos.

I ate one memorable lunch at the College at the height of the V-1 bombardment in the summer of 1944. My father and his professors were talking merrily about their plans for the expansion of the college to take care of the flood of students that would be pouring in as soon as the war was over. From time to time there was a momentary break in the conversation when the putt-putt-putt of an approaching V-1 could be heard in the distance. The talk and the jokes continued while the putt-putt-putt grew louder and louder until it seemed the beast must be directly overhead. Again there was a momentary break in the conversation when the putt-putt suddenly stopped, and the room was silent for the five seconds that it took the machine to descend to earth. Then an ear-splitting crash, and the conversation continued without a break until the next quiet putt-putt-putt could be heard in the distance. I was thinking lugubrious thoughts about the consequences that a direct hit on our dining room would have for the musical life of England. But such thoughts seemed to be far from the minds of my father and his colleagues. During the whole of our leisurely lunch, the subject of the V-1 bombardment was never once mentioned.

I used to talk a great deal with my father, especially during the early years of the war, about the morality of fighting and killing. At first I was a convinced pacifist and intended to become a conscientious objector. I agonized endlessly over the ethical line that had to be drawn between justifiable and unjustifiable participation in the war effort. My father listened patiently while I expounded my wavering principles and rationalized the latest shifts in my pacifist position. He said very little. My ethical doctrines grew more and more complicated as I was increasingly torn between my theoretical repudiation of national loyalties and my practical involvement in the life of a country fighting with considerable courage and good humor for its survival. For my father the issues were simple. He did not need to argue with me. He knew that actions speak louder than words. When he moved his bed into the college he made his position clear to everybody. When things were going badly in 1940, he said, "All we have to do is to behave halfway decently, and we shall soon have the whole world on our side." When he spoke of the whole world, he was probably thinking especially of the United States of America and of his own son.

Many years later I was reminded of these discussions between me and my father when I read the transcript of Oppenheimer's security hearing. The dramatic climax of the three-week hearing came near the end, when the physicist Edward Teller appeared as a witness for the prosecution and confronted Oppenheimer face to face. Teller was asked directly whether he considered Oppenheimer to be a security risk. He answered with carefully chosen words: "I thoroughly disagreed with him in numerous issues and his actions frankly appeared to me confused and complicated. To this extent I feel that I would like to see the vital interests of this country in hands which I understand better, and therefore trust more." These words describe rather accurately my father's attitude to my intellectual gyrations during the earlier part of the war. Oppenheimer, like me, was confused and complicated. He wanted to be on good terms with the Washington generals and to be a savior of humanity at the same time. Teller, like my father, was simple. He thought it was a dangerous illusion to imagine that we could save humanity by proclaiming high moral principles from a position of military weakness. He did his job as a scientist and bomb designer to keep America strong, and he left moral judgments concerning the use of our weapons to the American people and their elected representatives. Like my father, he believed that if we stayed strong and behaved decently the whole world would before long come to our side. His greatest mistake was his failure to foresee that a large section of the public would not consider his appearance at the Oppenheimer hearing to be decent behavior. Had Teller not appeared, the outcome of the hearing would almost certainly have been unaffected, and the moral force of Teller's position would not have been tainted.

The first time I met Teller was in March 1949, when I talked to the physicists at the University of Chicago about the radiation theories of Schwinger and Feynman. I diplomatically gave high praise to Schwinger and then explained why Feynman's methods were more useful and more illuminating. At the end of the lecture, the chairman called for questions from the audience. Teller asked the first question: "What would you think of a man who cried 'There is no God but Allah, and Mohammed is his prophet' and then at once drank down a great tankard of wine?" Since I remained speechless, Teller answered the question himself: "I would consider the man a very sensible fellow."

In 1949 the physics department at Chicago was second only to Cornell's in liveliness. Fermi and Teller in Chicago were like Bethe and Feynman at Cornell. Fermi the acknowledged leader, friendly and approachable but fundamentally serious. Teller bubbling over with ideas and jokes. Teller had done many interesting things in physics, but never the same thing for long. He seemed to do physics for fun rather than for glory. I took an instant liking to him.

I had been told in confidence by my friends at Cornell that Teller was deeply engaged in the American effort to build a hydrogen bomb. As a visiting foreigner I had no business to know about such things, but I was intensely curious to understand how a man with such a jovial and happy temperament could bring himself to work on the perfecting of engines of destruction even more fiendish than those we already possessed. In Chicago I found an opportunity to start an argument with him about politics. He revealed himself as an ardent supporter of the World Government movement, an organization which in those days promised salvation by means of a world government to be set up in the near future with or without the cooperation of the Soviet Union. Teller preached the gospel of world government with great charm and intelligence. I concluded my weekly report to my family with the words: "He is a good example of the saying that no man is so dangerous as an idealist."

Two years after my visit to Chicago, Teller and Ulam at Los Alamos made the crucial invention that changed the hydrogen bomb from a theoretical to a practical possibility. In 1949, before the Ulam-Teller invention, Oppenheimer had written of the hydrogen bomb, "I am not sure the miserable thing will work, nor that it can be gotten to a target except by ox-cart." After the invention, as Oppenheimer said at his trial, "From a technical point of view it was a sweet and lovely and beautiful job." Once the invention was made, in March 1951, it took the Los Alamos laboratory only twenty months of concentrated effort to build and explode a full-scale experimental bomb with a yield of ten million tons of TNT. A few years later, Teller published a historical account of the development of the bomb with the title "The Work of Many People," pointing out that he had received an excessively large share of both credit and blame for the bomb's existence. It was true that the bomb was very far from being the work of one man. Nevertheless, Teller had been the chief instigator and driving force, pushing indefatigably toward the bomb's realization, refusing to be discouraged by delays and difficulties, ever since the distant days of 1942 before Los Alamos began, through the wartime years and the years of frustration after 1945 when almost nobody would listen to him. He had thought longer and harder about hydrogen bombs than anybody else. It was no accident that he was the first to see how the things had to be built.

The invention and building of the hydrogen bomb in 1951–52 were hidden from public view. I was at the time at Cornell University, and all I knew about these matters was that Hans Bethe disappeared to Los Alamos for eight months at a stretch. That year I had to teach Hans's course in nuclear physics. Soon after Hans returned to Cornell, a gentleman from Washington came to visit with a briefcase chained to his wrist. The gentleman looked very uncomfortable standing at the physics department urinal with this massive briefcase dangling. No doubt the briefcase contained the results of the first hydrogen bomb test. Hans was preoccupied with things he could not talk about and seemed to have lost his zest for doing physics. It was a bad year at Cornell. One of the minor consequences of Hans's involvement with the hydrogen bomb was that I decided for the second time to move from Cornell to Princeton.

Two years later, when I was in Washington delivering the laundry to Oppenheimer's lawyer, I met Hans Bethe by chance in a hotel lobby. He was looking grimmer than I had ever seen him. I knew he had been testifying at Oppenheimer's trial. "Are the hearings going badly?" I asked. "Yes," said Hans, "but that is not the worst. I have just now had the most unpleasant conversation of my whole life. With Edward Teller." He did not say more, but the implications were clear. Teller had decided to testify against Oppenheimer. Hans had tried to dissuade him and failed.

This was a moment of tragedy for both Bethe and Teller. They had been close friends for many years, since long before the war. Their temperaments and abilities complemented each other wonderfully, Teller with his high spirits and free-ranging imagination, Bethe with his seriousness and powerful common sense. Before Bethe married, he was so often a guest in the Teller home that he became almost one of the family. In April 1954 that was all over. There could be no real reconciliation. Bethe had lost one of his oldest friends. But Teller had lost more. Teller, by lending his voice to the cause of Oppenheimer's enemies, had lost not only the friendship but the respect of many of his colleagues. He was portrayed by newspaper writers and cartoonists as a Judas, a man who had betrayed his leader for the sake of personal gain. A careful reading of his testimony at the trial shows that he intended no personal betrayal. He wanted only to destroy Oppenheimer's political power, not to damage Oppenheimer personally. But the mood of that time made such fine distinctions meaningless. In the eyes of the majority of scientists and academic people, Oppenheimer's trial was simply a campaign led by a group of paranoid patriots who were trying to silence opposition to their policies by a personal attack on their most visible opponent. By joining the campaign, no matter what he said and no matter why he said it, Teller made himself an object of hatred and distrust to a whole generation of young people. He wounded himself more grievously than he wounded Oppenheimer. Like Oppenheimer before him, Teller, too, had been seduced by the Demon at the summit of F6. The abbot in the monastery had foretold their fate in his warning to M.F.:

As long as the world endures, there must be order, there must be government: but woe to the governors, for, by the very operation of their duty, however excellent, they themselves are destroyed. For you can only rule men by appealing to their fear and their lust; government requires the exercise of the human will: and the human will is from the Demon."

Nuclear explosives have a glitter more seductive than gold to those who play with them. To command nature to release in a pint pot the energy that fuels the stars, to lift by pure thought a million tons of rock into the sky, these are exercises of the human will that produce an illusion of illimitable power. Oppenheimer and Teller each came to perform these exercises of the human will for good and honest reasons. Oppenheimer was driven to build atomic bombs by the fear that if he did not seize this power, Hitler would seize it first. Teller was driven to build hydrogen bombs by the fear that Stalin would use this power to rule the world. Oppenheimer, being Jewish, had good reason to fear Hitler. Teller, being Hungarian, had good reason to fear Stalin. But each of them, having achieved his technical objective, wanted more. Each of them was led by his Demon to seek political as well as technical power. Each of them became convinced that he must have political power to ensure that the direction of the enterprise he had created should not fall into hands that he considered irresponsible. In the end, each of them was irrevocably committed to exercises of the human will in the political as well as the technical sphere. And so each of them in his own way came to grief.

While the secret battles over the hydrogen bomb were raging, I was quietly raising babies and continuing to think about electrons. I spent several summers at the University of California in Berkeley, teaching summer school courses and working with Charles Kittel on the theory of electrons in metals. Metals conduct electricity because their electrons are not attached to individual atoms but are free to move around independently. To understand a metal it is not enough to understand the behavior of electrons one at a time. One must deal with electrons in large numbers, and this raises new problems. It turns out that the methods that Schwinger and Feynman invented for describing individual electrons can be adapted to give a good account of electrons in metals. I made a beginning with the adaptation.

In the summer of 1955 I rented a big house in Berkeley for my growing family. That summer I was working happily with Charles Kittel's group of solid-state physicists, trying to understand spin waves. Spin waves are waves of magnetization that can travel through a solid magnet as ocean waves travel through water. Tickle a magnet with a rapidly varying magnetic field and the spin waves start running. From the way they run and the way they die down, experimenters obtain detailed information about the atomic structure of the magnet. I spent the summer struggling to put together an exact mathematical description of spin waves rolling on the sea of atoms. It is easy to describe a magnet as a collection of atoms. It is easy to describe it as a collection of spin waves. The difficult problem is to connect these two partial pictures together in a coherent scheme that includes both. I made some progress with this problem but did not solve it completely. It is still not completely solved, twenty-two years later.

The house that we rented for the summer stood on the hill overlooking the Berkeley campus. It was a magnificent house with a magnificent view, and above it the hillside was still wild. We could walk from the house into eucalyptus woods where our children liked to play. One Sunday morning we went for a walk up the hill, leaving the house open as usual. When we came back through the trees to the house, we heard a strange sound coming through the open door. The children stopped their chatter and we all stood outside the door and listened. It was my old friend from long ago, Bach's Prelude No. 8 in E-flat minor. Superbly played. Played just the way my father used to play it. For a moment I was completely disoriented. I thought: What the devil is my father doing here in California?

We stood in front of our Berkeley house and listened to that prelude. Whoever was playing it, he was putting into it his whole heart and soul. The sound floated up to us like a chorus of mourning from the depths, as if the spirits in the underworld were dancing to a slow pavane. We waited until the music came to an end and then walked in. There, sitting at the piano, was Edward Teller. We asked him to go on playing, but he excused himself. He said he had come to invite us to a party at his house and had happened to see that fine piano begging to be played. We accepted the invitation and he went on his way. That was the first time I had spoken with him since our encounter six years earlier in Chicago. I decided that no matter what the judgment of history upon this man might be, I had no cause to consider him my enemy.

9

Little Red Schoolhouse

Eddington the astronomer, in the book *New Pathways in Science*, which I read as a boy in Winchester, not only warned us against nuclear bombs but promised us nuclear power stations. Here is the happier side of his vision of the future:

> We build a great generating station of, say, a hundred thousand kilowatts capacity, and surround it with wharves and sidings where load after load of fuel is brought to feed the monster. My vision is that some day these fuel arrangements will no longer be needed; instead of pampering the appetite of the engine with delicacies like coal and oil, we shall induce it to work on a plain diet of subatomic energy. If that day ever arrives, the barges, the trucks, the cranes will disappear, and the year's supply of fuel for the power-station will be carried in in a tea-cup.

This vision had always remained vivid in my mind, together with the warning against the military use of subatomic energy which appears a few pages later in the book. Eddington used the word "subatomic" to describe what we now call nuclear or atomic energy. We all knew even in 1937 that the world would soon run out of coal and oil. The possible availability of nuclear energy to satisfy the peaceful needs of mankind was one of the few hopeful prospects in a dark period of history.

In August 1955, while I was quietly working on spin waves in Berkeley, a mammoth international conference on the peaceful uses of atomic energy was held in Geneva under the auspices of the United Nations. This was a decisive moment in the development of nuclear energy. American and British and French and Canadian and

Russian scientists, who had been building nuclear reactors in isolation and secrecy, were able for the first time to meet one another and discuss their work with considerable freedom. Masses of hitherto secret documents were presented openly to the conference, making available to scientists of all countries almost all the basic scientific facts about the fission of uranium and plutonium and a large fraction of the engineering information that was needed for the building of commercial reactors. A spirit of general euphoria prevailed. Innumerable speeches proclaimed the birth of a new era of international cooperation, the conversion of intellectual and material resources away from weapon building into the beneficent pursuit of peaceful nuclear power, and so on and so on. Some part of what was said in these speeches was true. The conference opened channels of communication between the technical communities in all countries, and the personal contacts which were established in 1955 have been successfully maintained ever since. To some small extent, the habit of openness in international discussions of peaceful nuclear technology has spread into the more delicate areas of weaponry and politics. The high hopes raised in Geneva in 1955 have not proved entirely illusory.

The technical preparations for the Geneva meeting were made by an international group of seventeen scientific secretaries. The scientific secretaries worked in New York for several months, driving hard bargains on behalf of their governments, making sure that each participating country would reveal a fair share of its secrets and receive a fair share of the limelight. They worked in obscurity and waded through vast quantities of paper. The success of the conference was entirely due to their efforts. One of the two Americans in the group of seventeen was Frederic de Hoffmann, a thirty-year-old physicist then employed as a nuclear expert by the Convair Division of the General Dynamics Corporation in San Diego, California.

As soon as the Geneva meeting was over, Freddy de Hoffmann decided the time had come to give the commercial development of nuclear energy a serious push. For the first time it would be possible to build reactors and sell them on the open market, free from the bureaucratic miseries of secrecy. He persuaded the top management of the General Dynamics Corporation to set up a new division called General Atomic, with himself as president. General Atomic began its life at the beginning of 1956 with no buildings, no equipment and no staff. Freddy rented a little red schoolhouse that had been abandoned as obsolete by the San Diego public school system. He proposed to move into the schoolhouse and begin designing reactors there in June.

Freddy had been at Los Alamos with Edward Teller in 1951 and had made some of the crucial calculations leading to the invention of the hydrogen bomb. He invited Teller to join him in the schoolhouse for the summer of 1956. Teller accepted with enthusiasm. He knew that he and Freddy could work well together, and he shared Freddy's strong desire to get away from bombs for a while and do something constructive with nuclear energy.

Freddy also invited thirty or forty other people to spend the summer in the schoolhouse, most of them people who had been involved with nuclear energy in one way or another, as physicists, chemists or engineers. Robert Charpie, even younger than Freddy, had been the other American in the group of scientific secretaries of the Geneva meeting. Ted Taylor came directly from Los Alamos, where he had been the pioneer of a new art form, the design of small efficient bombs that could be squeezed into tight spaces. For some reason, although I had never had anything to do with nuclear energy and was not even an American citizen, I was also on Freddy's list. Probably this was a result of my encounter with Teller the previous summer. Freddy promised me a chance to work with Teller. I accepted the invitation gladly. I had no idea whether I would be successful as a reactor designer, but at least I would give it a try. For nineteen years I had been waiting for this opportunity to make Eddington's dream come true.

Freddy de Hoffmann was my first encounter with the world of Big Business. I had never before met anybody with the authority to make decisions so quickly and with so little fuss. I found it remarkable that this authority was given to somebody so young. Freddy handled his power lightly. He was good-humored, and willing to listen and learn. He always seemed to have time to spare.

We assembled in June in the schoolhouse, and Freddy told us his plan of work. Every morning there would be three hours of lectures. The people who were already expert in some area of reactor technology would lecture and the others would learn. So at the end of the summer we would all be experts. Meanwhile we would spend the afternoons divided into working groups to invent new kinds of reactors. Our primary job was to find out whether there was any specific type of reactor that looked promising as a commercial venture for General Atomic to build and sell.

The lectures were excellent. They were especially good for me, coming into the reactor business from a position of total ignorance. But even the established experts learned a lot from each other. The physicists who knew everything that was to be known about the physics of reactors learned about the details of the chemistry and engineering. The chemists and engineers learned about the physics. Within a few weeks we were all able to understand each other's problems.

The afternoon sessions quickly crystallized into three working groups, with the titles "Safe Reactor," "Test Reactor" and "Ship Reactor." These were considered to be the three main areas where an immediate market for civilian reactors might exist. In retrospect it seems strange that electricity-producing power reactors were not on our list. Freddy knew that General Atomic must ultimately get into the power reactor business, but he wanted the company to begin with something smaller and simpler to gain experience. The ship reactor was intended to be a nuclear engine for a merchant ship, and the test reactor was intended to be a small reactor with a very high neutron flux which could be used for the testing of component parts of power reactors. Both these reactors would be competing directly with existing reactors that had already been developed for the Navy and the Atomic Energy Commission. Both of them were designed during the summer and then abandoned when Freddy concluded that they had no commercial future. The safe reactor was the only product of our little red schoolhouse which actually got built.

The safe reactor was Teller's idea, and he took charge of it from the beginning. He saw clearly that the problem of safety would be decisive for the long-range future of civilian reactors. If reactors were unsafe, nobody in the long run would want to use them. He told Freddy that the best way for General Atomic to break quickly into the reactor market was to build a reactor that was demonstrably safer than anybody else's. He defined the task of the safe reactor group in the following way: The group was to design a reactor so safe that it could be given to a bunch of high school children to play with, without any fear that they would get hurt. This objective seemed to me to make a great deal of sense. I joined the safe reactor group and spent the next two months with Teller fighting our way through to a satisfactory solution of his problem.

Working with Teller was as exciting as I had imagined it would be. Almost every day he came to the schoolhouse with some hare-brained new idea. Some of his ideas were brilliant, some were practical, and a few were brilliant and practical. I used his ideas as starting points for a more systematic analysis of the problem. His intuition and my mathematics fitted together in the design of the safe reactor just as Dick Feynman's intuition and my mathematics had fitted together in the understanding of the electron. I fought with Teller as I had fought with Feynman, demolishing his wilder schemes and squeezing his intuitions down into equations. Out of our fierce disagreements the shape of the safe reactor gradually emerged. Of course I was not alone with Teller as I had been with Feynman. The safe reactor group was a team of ten people. Teller and I did most of the shouting, while the chemists and engineers in the group did most of the real work.

Reactors are controlled by long metal rods containing substances such as boron and cadmium, which absorb neutrons strongly. When you want to make the reactor run faster, you pull the control rods a little way out of the reactor core. When you want to shut the reactor down, you push the control rods all the way in. The first rule in operating a reactor is that you do not suddenly yank the control rods out of a shut-down reactor. The result of suddenly pulling out the control rods would in most cases be a catastrophic accident, including as one of its minor consequences the death of the idiot who pulled the rods. All large reactors are therefore built with automatic control systems which make it impossible to pull the rods out suddenly. These reactors possess "engineered safety," which means that a catastrophic accident is theoretically possible but is prevented by the way the control system is designed. For Teller, engineered safety was not good enough. He asked us to design a reactor with "inherent safety," meaning that its safety must be guaranteed by the laws of nature and not merely by the details of its engineering. It must be safe even in the hands of an idiot clever enough to by-pass the entire control system and blow out the control rods with dynamite. Stated more precisely, Teller's ground rule for the safe reactor was that if it was started from its shut-down condition and all its control rods instantaneously removed, it would settle down to a steady level of operation without melting any of its fuel.

One of the first steps toward the design of the safe reactor was to introduce an idea called the "warm neutron principle," which says that warm neutrons are less easily captured than cold neutrons and are less effective in causing uranium atoms to fission. The neutrons in a water-cooled reactor are slowed down by collisions with hydrogen atoms and end up with roughly the same temperature as the hydrogen in whatever place they happen to be. In an ordinary water-cooled reactor, after the postulated idiot has blown out the control rods, the fuel will be growing rapidly hot but the water will still be cold, with the result that the neutrons remain cold and their effectiveness in causing fission is undiminished, and therefore the fuel continues to grow hotter until it finally melts or vaporizes. But suppose instead that the reactor was designed with only half of the hydrogen in the cooling water and the other half of the hydrogen mixed into the solid structure of the fuel rods. In this case, when the idiot yanks out the control rods, the fuel will grow hot and with it the hydrogen in the fuel rods, while the hydrogen in the water remains cold. The result is then that the neutrons inside the fuel rods are warmer than the neutrons in the water. The warm neutrons cause less fission and escape more easily into the water to be cooled and captured, and the reactor automatically stabilizes itself within a few thousandths of a second, much faster than any mechanical safety switch could hope to operate. So the reactor carrying half of its hydrogen in its fuel rods is inherently safe.

There were many practical difficulties to be overcome before these ideas could be embodied in functioning hardware. The greatest contribution to overcoming the practical difficulties was made by Massoud Simnad, an Iranian metallurgist who discovered how to make fuel rods containing high concentrations of hydrogen. He made the rods out of an alloy of uranium hydride with zirconium hydride. He found the right proportions of these ingredients to mix together and the right way to cook them. When the fuel rods emerged from Massoud's oven, they looked like black, hard, shiny metal, as tough and as corrosion-resistant as good stainless steel.

After we had understood the physics of the safe reactor and the chemistry of its fuel rods, many questions still remained to be answered. Who would want to buy such a reactor? What would they use it for? How powerful should it be? How much should it cost? Teller insisted from the beginning that it should not be just a toy for reactor experts to play with. It must be not only safe, but also powerful enough to do something useful. What could it do?

The most plausible use for a reactor of this kind would be to produce short-lived radioactive isotopes for medical research and diagnosis. When radioactive isotopes are used as biochemical tracers to study malfunctions in living people, it is much better to use isotopes that decay within a few minutes or hours so that they are gone as soon as the observation is over. The disadvantage of short-lived isotopes is that they cannot be shipped from one place to another. They must be made where they are used. So our safe reactor might come in handy for a big research hospital or medical center that wanted to produce its own isotopes. We calculated that for this purpose a power level of one megawatt would be generally adequate. The other uses that we envisaged for our reactor were for training students in nuclear engineering departments of universities, and for doing research in metallurgy and solid-state physics using beams of neutrons to explore the structure of matter. If the reactor was used for neutron beam research, a power of one megawatt would be rather low, and so we also designed a high-powered version that could be run at ten megawatts. Freddy named the safe reactor TRIGA, the letters standing for Training, Research and Isotopes, General Atomic.

In September the summer's work in San Diego was coming to an end and I took a bus ride to Tijuana in Mexico to buy presents for my family. As I was walking through Tijuana after dark, a small dog ran up to me from behind and bit me in the leg. Tijuana was so overrun with sickly and mangy dogs that there was no chance whatever of catching and identifying the animal that bit me. So I went to a clinic in La Jolla every day for fourteen days to take the Pasteur treatment against rabies. The doctor who gave me the injections impressed on me forcefully the fact that the treatment itself was risky, causing in one case out of six hundred an allergic encephalitis which was almost as fatal as rabies. He told me to figure the odds carefully before beginning the treatment. I decided to take the shots, and I was consequently under some emotional strain for the last two weeks of the summer. Edward Teller was extremely helpful. He had in his youth in Budapest lost a foot in a streetcar accident, and he knew how to give effective moral support in a situation of this kind. In

Berkeley I had decided not to consider him an enemy. In San Diego he became a lifelong friend.

After Teller and I and the rest of the summer visitors departed, the few people who remained at General Atomic undertook the job of turning our preliminary sketches of the Triga into a working reactor. The final design was worked out by Ted Taylor, Stan Koutz and Andrew McReynolds. It took less than three years from Teller's original proposal in the summer of 1956 for the first batch of Trigas to be built, licensed and sold. The basic price was a hundred and forty-four thousand dollars, not including the building. The Trigas sold well and have continued to sell ever since. The last time I checked the total, sixty had been sold. It is one of the very few reactors that made money for the company which built it.

In June 1959, all the people who had worked in the schoolhouse to get General Atomic started were invited back to attend the official dedication ceremonies of the General Atomic Laboratories. The change in three years was startling. Instead of a rented schoolhouse, Freddy now had a magnificent set of permanent buildings constructed in a modernistic style on a mesa on the northern edge of San Diego. He had well-equipped laboratories and machine shops, with a staff already growing into the hundreds. In one of the buildings was the prototype Triga, fully licensed and ready to perform for prospective customers. Freddy had persuaded Niels Bohr himself, by common consent the greatest living physicist after the death of Einstein, to come from Copenhagen to preside over the dedication.

The climax of the dedication ceremony was a demonstration of the capabilities of the Triga. Freddy had attached to the speaker's podium a switch and a large illuminated dial. At the end of his speech, Niels Bohr pressed the switch and a muffled hiss was heard from the direction of the Triga building. The noise came from the sudden release of compressed air that was used to pull the control rods at high speed out of the Triga core. The pointer on the large dial, which was graduated to show the power output of the Triga in megawatts, swung over instantaneously to 1500 megawatts and then quickly subsided to half a megawatt. The demonstration was over. It had been rehearsed many times before, to make sure there would be no unpleasant surprises. The little reactor did in fact run at a rate of 1500 megawatts for a few thousandths of a second before its warm neutrons brought it under control. After the ceremony we went and saw it sitting quietly at the bottom of its pool of cooling water. Here it was. It was hard to believe. How could one believe that nature would pay attention to all the theoretical arguments and calculations that we had fought over in the schoolhouse three years earlier? But here was the proof. Warm neutrons really worked.

In the evening there was a picnic supper on the beach, with Freddy and Niels Bohr and various other dignitaries. After eating, Bohr became restless. It was his habit to walk and talk. All his life he had been walking and talking, usually with a single listener who could concentrate his full attention upon Bohr's convoluted sentences and indistinct voice. That evening he wanted to talk about the future of atomic energy. He signaled to me to come with him, and we walked together up and down the beach. I was delighted to be so honored. I thought of the abbot in the monastery at the foot of F6, and I wondered whether it would now be my turn to look into the crystal ball. Bohr told me that we now had another great opportunity to gain the confidence of the Russians by talking with them openly about all aspects of nuclear energy. The first opportunity to do this had been missed in 1944, when Bohr spoke with both Churchill and Roosevelt and failed to persuade them that the only way to avoid a disastrous nuclear arms race was to deal with the Russians openly before the war ended. Bohr talked on and on about his conversations with Churchill and Roosevelt, conversations of the highest historical importance which were, alas, never recorded. I clutched at every word as best I could. But Bohr's voice was at the best of times barely audible. There on the beach, each time he came to a particularly crucial point of his confrontations with Churchill and Roosevelt, his voice seemed to sink lower and lower until it was utterly lost in the ebb and flow of the waves. That night the abbot's crystal ball was cloudy.

For Freddy, the Triga was only a beginning. He knew that General Atomic's survival would in the end depend on its ability to build and sell full-scale power reactors. Already in 1959 the major part of the laboratory's efforts were devoted to the development of a power reactor. Freddy had decided to stake his future on a particular type of power reactor, the High Temperature Graphite Reactor or HTGR. All of us who were involved with General Atomic supported this decision. It was a big gamble, and it ultimately failed. But I still think Freddy's decision was right. If he had been as lucky with the HTGR as he was with the Triga, it would have paid off handsomely for General Atomic, and the whole nuclear industry of the United States would be in much better shape than it now is. It is impossible to make real progress in technology without gambling. And the trouble with gambling is that you do not always win.

The HTGR was competing directly with the light-water power reactors which have from the beginning monopolized the United States nuclear power industry. Neither HTGR nor light-water reactors are inherently safe in the sense that the Triga is safe. Both depend on engineered safety systems to push in the control rods and shut down the nuclear reaction in case of any trouble. Both have enough residual radioactivity to vaporize the core and cause a major accident if the cooling of the core is not continued after shutdown. The main difference between the two reactors is that the HTGR has a much bigger core for the same output of heat. The HTGR core has such a great capacity for soaking up heat that it will take many hours to reach the melting point after a shutdown, even if there is a complete failure of emergency cooling systems. A light-water power reactor core will melt in a few minutes under the same conditions. The worst conceivable HTGR accident would be an exceedingly messy affair, but it would be definitely less violent and less unmanageable than a comparable accident in a light-water reactor. In this sense the HTGR is a fundamentally safer system.

The HTGR is not only safer than a light-water reactor but also more efficient in its use of fuel. These are its two great advantages. It has two great disadvantages: It is more expensive to build, and it has more difficulty with controlling the leakage of small quantities of radioactive fission products during normal operation. Freddy gambled on the expectation that superior safety and efficiency would in the long run cause the world to turn to the HTGR for electric power. He may well turn out to have been right, but the long run was too long for his company. In the short run, the disadvantages of capital cost and of complexity of the leakage containment system stopped him from breaking into the market. He sold only two HTGRs and never went into production with a full-scale model. Finally, in the late 1970s the political uncertainties surrounding nuclear power made the outlook for the HTGR seem commercially hopeless. General Atomic canceled its contracts with its few remaining HTGR customers and announced that it was no longer in the fission power reactor business. Several years earlier, Freddy had moved across the street from General Atomic to become president of the Salk Institute for Biological Studies. General Atomic still continues to build and sell Trigás and to support an active program of research in controlled fusion. No longer is nuclear fission power a promising new frontier for young scientists and forward-looking businessmen.

What went wrong with nuclear power? When Freddy invited me to work on reactors in 1956, I jumped at the opportunity to apply my talents to this great enterprise of bringing cheap and unlimited energy to mankind. Edward Teller and the other inhabitants of the schoolhouse all felt the same way about it. Finally we were learning how to put nuclear energy to better use than building bombs. Finally we were going to do some good with nuclear energy. Finally we were going to supply the world with so much energy that human drudgery and poverty would be abolished. What went wrong with our dreams?

There is no simple answer to this question. Many historical forces conspired to make the development of nuclear energy more troublesome and more costly than we had expected. If we had been wiser, we might have foreseen that after thirty years of unfulfilled promises a new generation of young people and of political leaders would arise who regard nuclear energy as a trap from which it is their mission to liberate us. It is only natural that the dreams of thirty years ago should not appeal to the young people of today. They need new visions to keep them moving ahead. It is easy to understand in a general way why the political atmosphere surrounding nuclear energy has changed so markedly for the worse since the days of the little red schoolhouse. But I believe there is a more specific explanation for many of the troubles which now beset the nuclear power industry. This is the fact that within the industry itself, the spirit of the schoolhouse did not prevail.

The fundamental problem of the nuclear power industry is not reactor safety, not waste disposal, not the dangers of nuclear proliferation, real though all these problems are. The fundamental problem of the industry is that nobody any longer has any fun building reactors. It is inconceivable under present conditions that a group of enthusiasts could assemble in a schoolhouse and design, build, test, license and sell a reactor within three years. Sometime between 1960 and 1970, the fun went out of the business. The adventurers, the experimenters, the inventors, were driven out, and the accountants and managers took control. Not only in private industry but also in the government laboratories, at Los Alamos, Livermore, Oak Ridge and Argonne, the groups of bright young people who used to build and invent and experiment with a great variety of reactors were disbanded. The accountants and managers decided that it was not cost effective to let bright people play with weird reactors. So the weird reactors disappeared and with them the chance of any radical improvement beyond our existing systems. We are left with a very small number of reactor types in operation, each of them frozen into a huge bureaucratic organization that makes any substantial change impossible, each of them in various ways technically unsatisfactory, each of them less safe than many possible alternative designs which have been discarded. Nobody builds reactors for fun any more. The spirit of the little red schoolhouse is dead. That, in my opinion, is what went wrong with nuclear power.

When my father was a young man, he used to travel around Europe on a motorcycle. Sixty years before Robert Pirsig, he learned to appreciate the art of motorcycle maintenance and the virtue of a technology based upon respect for quality. He sometimes came to villages where no motorcycle had been before. In those days every rider was his own repairman. Riders and manufacturers were together engaged in trying out a huge variety of different models, learning by trial and error which designs were rugged and practical and which were not. It took thousands of attempts, most of which ended in failure, to evolve the few types of motorcycle that are now on the roads. The evolution of motorcycles was a Darwinian process of the survival of the fittest. That is why the modern motorcycle is efficient and reliable.

Contrast this story of the motorcycle with the history of commercial nuclear power. In the worldwide effort to develop an economical nuclear power station, less than a hundred different types of reactor have been operated. The number of different types under development grows constantly smaller, as the political authorities in various countries eliminate the riskier ventures for reasons of economy. There now exist only about ten types of nuclear power station that have any hope of survival, and it is impossible under present conditions for any radically new type to receive a fair trial. This is the fundamental reason why nuclear power plants are not as successful as motorcycles. We did not have the patience to try out a thousand different designs, and so the really good reactors were never invented. Perhaps it is true in technology as it is in biological evolution that wastage is the key to efficiency. In both domains, small creatures evolve more easily than big ones. Birds evolved while their cousins the dinosaurs died.

Is there any hope for the future of nuclear power? Of course there is. The future is unpredictable. Political moods and fashions change fast. One fact that will not change is that mankind will need enormous quantities of energy after the oil runs out. Mankind will see to it that the energy is produced, one way or another. When that day comes, people will need nuclear power reactors cheaper and safer than those we are now building. Perhaps our managers and accountants will then have the wisdom to assemble a group of enthusiasts in a little red schoolhouse and give them some freedom to tinker around.

10

Saturn by 1970

The beginning of the space age can be dated rather precisely, to June 5, 1927, when nine young men meeting in a restaurant in the German town of Breslau (now the Polish town of Wroclaw) founded the Verein für Raumschiffahrt. The German name means Space-Travel Society and is generally abbreviated to VfR. The VfR existed for six years before Hitler put an end to it, and in those six years it carried through the basic engineering development of liquid-fueled rockets without any help from the German government. This was the first romantic age in the history of space flight. The VfR was an organization without any organization. It depended entirely upon the initiative and devotion of individual members. Wernher von Braun joined the society as an eighteen-year-old student in 1930 and played an active part in it for the last three years of its existence. In a strange way, the last desperate years of the Weimar Republic produced at the same time the splendid flowering of pure physics in Germany and the legendary achievements of the VfR, as if the young Germans of that time were driven to make their highest creative efforts by the economic and social disintegration which surrounded them. The VfR was also lucky to have among its founding members a historian, who was something of a poet as well as a first-rate engineer. By his writings Willy Ley saved the legends of the VfR from oblivion, as Chaucer saved the tales of the pilgrims who rode with him to Canterbury.

Willy Ley was twenty-one when he helped to found the VfR, and twenty-seven when the VfR died. In his book *Rockets, Missiles and Space Travel,* he describes the drama of the first successful VfR rocket flight. "Our rocket testing-ground had grown very beautiful with the coming of Spring. The hilly part was covered with the young green of pine shoots and new birch leaves, the depressions between the hills were full of young willows. Crickets sang in the high grass and frogs croaked somewhere in the distance. . . . But the beast flew! Went up like an elevator, very slowly, to twenty meters. Then it fell down and broke a leg." That was May 10, 1931, on a swampy piece of land within the city limits of Berlin. In one year of frenzied work the remaining difficulties were overcome, and by the summer of 1932 the VfR rockets were flying reliably to heights of one or two kilometers.

A year later, Hitler was in power, and all the journals, books, correspondence and records of the VfR were seized by the Gestapo. In 1933 the era of poets and amateurs was over and the era of the professionals had begun. A member of the VfR who worked for the Siemens company in Berlin overheard one of the company managers telephoning a friend in the War Ministry: "Now I've all the rocket people safely on ice around here and can watch what they are doing." The development of rocketry was taken over by the military, who set up their research and testing organization at a remote site called Peenemünde on the Baltic Sea, with big money, big bureaucracy, and twenty thousand employees. Von Braun was installed there as technical director. The result of this great professional effort was what might have been expected, a technically brilliant device, the V-2 rocket, which made no economic or military sense. I became aware of the success of the Peenemünde project in the fall of 1944, after the V-1 bombardment of London had ended, when I heard the occasional bang of a V-2 warhead exploding. At night, when the city was quiet, you could hear after the bang the whining sound of the rocket's supersonic descent. At that time in London, those of us who were seriously engaged in the war were very grateful to Wernher von Braun. We knew that each V-2 cost as much to produce as a high-performance fighter airplane. We knew that the German forces on the fighting fronts were in desperate need of airplanes, and that the V-2 rockets were doing us no military damage. From our point of view, the effect of the V-2 program was almost as good as if Hitler had adopted a policy of unilateral disarmament. Unilateral disarmament had certainly not been the intention of the military leaders who set up the Peenemünde organization. This is an extreme example of the stupidities which often occur when bureaucracy takes control of scientific projects. Such stupidities are by no means an exclusively German phenomenon.

My own involvement with the exploration of space began early in the year 1958. Freddy de Hoffmann passed through Princeton and told me the latest news of the operational trials of the prototype Triga. "By the way," he said, "Ted Taylor has a crazy idea for a nuclear spaceship and he wants you to come out to San Diego and look at it." I went. This was the beginning of Project Orion.

After the summer in the schoolhouse, Ted Taylor had decided to move permanently from Los Alamos to General Atomic. He helped Freddy organize the new laboratories, and he supervised the design and construction of the prototype Triga reactor. But his head was still full of the elegant little bombs he had been designing at Los Alamos. During idle moments he began thinking again about an idea that had been suggested some years before by Ulam at Los Alamos. Could one not use these elegant little bombs to drive an elegant little spaceship around the solar system?

Ted was two years younger than I, and completely unknown to the public. He was neither a scientific genius like Dick Feynman nor a flamboyant personality like Freddy de Hoffmann. He was quiet and unhurried. Since those days he has become an important public figure, and John McPhee has written a book describing his life and achievements. I do not know how it happened that I saw the greatness in him from the beginning. Outwardly, he looked like an ordinary American Westerner, with a philosophical wife and four rowdy children. Inside, there was a tremendous detachment, imagination and stubbornness. Nobody but Ted could have led Project Orion and kept his undisciplined band of followers working on it with a passionate prodigality through good times and bad for five long years.

In the summer of 1957 the first Russian Sputnik went up. A few months later Wernher von Braun, with the resources of the United States Army now at his disposal, launched his first satellite in reply. The battle of the giants had begun. Big and ponderous organizations were in command on both sides. People in our government were already speaking about a project to land men on the moon with huge conventional rockets, a project which would take ten years and twenty billion dollars to complete. Ted was interested in going into space but was repelled by the billion-dollar style of the big government organizations. He wanted to recapture the style and spirit of the VfR. And for a short time he succeeded.

Ted started out with three basic beliefs. First, the conventional Von Braun approach to space travel using chemical rockets would soon run into a dead end, since manned flights going farther than the moon would become absurdly expensive. Second, the key to interplanetary flight must be the use of nuclear fuel, which carries in each pound a million times as much energy as chemical fuel. Third, a small group of people with daring and imagination could design a nuclear spaceship which would be cheaper and enormously more capable than the best chemical rocket. So Ted set to work in the spring of 1958 to create his own VfR. Freddy allowed him to use the facilities of General Atomic and gave him a small amount of company money to get started. I agreed to come and work on Orion full time for the academic year 1958–59. We intended to build a spaceship which would be simple, rugged, and capable of carrying large payloads cheaply all over the solar system. Our slogan for the project was "Saturn by 1970."

Already in 1958 we could see that Von Braun's moon ships, the ships that were to be used for the Apollo voyages to the moon ten years later, would cost too much and do too little. In many ways the Apollo ships were like the V-2 rockets. Both were brainchildren of Wernher von Braun. Both were magnificent technological achievements. Both were far too expensive for the limited job they were designed to do. The Apollo ships were superbly successful in taking men for short trips to the moon, and they looked beautiful on television. But as soon as mankind became tired of this particular spectacle, the Apollo ships became as obsolete as the V-2. There was nothing else that they could do.

Ted and I felt from the beginning that space travel must become cheap before it could have a liberating influence upon human affairs. So long as it costs hundreds of millions of dollars to send three men to the moon, space travel will be a luxury that only big governments can afford. And high costs make it almost impossible to innovate, to modify the propulsion system, or to adapt it to a variety of purposes. Project Orion proposed to lift large payloads from the ground into orbit around the earth at a cost of a few dollars a pound, about a hundred times cheaper than chemical rockets can do it. We were confident that once we had achieved cheap transportation into orbit, interplanetary missions would soon follow. We sketched a twelve-year flight program ending with large manned expeditions, to Mars in 1968 and to the satellites of Jupiter and Saturn in 1970. The costs of our program added up to about one hundred million dollars a year. Of course none of the professional accountants believed our cost estimates. Probably they were right. But I am not sure. For Ted and me the words "Saturn by 1970" were not just an idle boast. We really believed we could do it if we were given the chance. We took turns looking at Jupiter and Saturn through a little telescope that Ted kept in his garden. In our imagination we were zooming under the arch of Saturn's rings to make the last braking maneuver before landing on the satellite Enceladus. Enceladus was our favorite landing place because it is one of the places in the solar system where water is certain to be found in abundance. There we could replenish our supplies of water for the homeward voyage, and perhaps also do a little hydroponic farming and raise a crop of fresh vegetables.

In July 1958, when Project Orion was formally established, I wrote a document called "A Space Traveler's Manifesto" to describe to the world what we were doing and why. This is what it said:

The American government has announced that we are thinking about the design of a space-ship to be driven by atomic bombs. . . . It is my belief that this scheme alone, of the many space-ship schemes that are under consideration, can lead to a ship adequate to the real magnitude of the task of exploring the Solar System. We are fortunate in that the government has advised us to go straight ahead for the long-range scientific objectives of inter-planetary travel, and to disregard possible military uses of our propulsion system. . . .

From my childhood it has been my conviction that men would reach the planets in my lifetime, and that I should help in the enterprise. If I try to rationalize this conviction, I suppose it rests on two beliefs, one scientific and one political:

(1) There are more things in heaven and earth than are dreamed of in our present-day science. And we shall only find out what they are if we go out and look for them.

(2) It is in the long run essential to the growth of any new and high civilization that small groups of people can escape from their neighbors and from their governments, to go and live as they please in the wilderness. A truly isolated, small, and creative society will never again be possible on this planet.

To these two articles of faith I have now to add a third:

(3) We have for the first time imagined a way to use the huge stockpiles of our bombs for better purpose than for murdering people. Our purpose, and our belief, is that the bombs which killed and maimed at Hiroshima and Nagasaki shall one day open the skies to man.

We worked together for a year, from summer 1958 to fall 1959, as full of enthusiasm as the VfR pioneers in their great year from 1931 to 1932. We, too, were working in a hurry, knowing that we had little time before the fall of night. We knew that the government must soon decide whether to put its main effort into chemical or into nuclear propulsion, and if we were not ready with a workable design the choice would inevitably go against us.

We worked simultaneously at four different levels: theoretical physics calculations, experiments with high-velocity gas jets, engineering design of full-scale ships, and flight testing of models. At the beginning we had no specialists. Just as in the VfR, everybody did a little of everything. Later we became slightly bureaucratic and divided ourselves into physicists and engineers.

The most beautiful part of the project was the flight testing. We built model ships which propelled themselves with chemical high-explosive charges instead of with nuclear bombs. One of our team was Jerry Astl, a Czech refugee scientist who was an artist with high explosives. He knew how to build complicated high-explosive devices with elaborate fusing and timing systems, and they almost always worked. He had learned his trade in the Czech underground during World War II.

We had our test site on Point Loma, a steep peninsula of land which sticks out into the Pacific Ocean west of the city of San Diego. The land belongs to the United States Navy and has been saved from the cancer of real estate development which has spoiled the Pacific coast north and south of it. Our site had on it only a small rocket test stand, long ago abandoned by the Navy. There was no other sign of man's presence. All around us was the untouched hillside, covered with green shrubs and flowering cactus. Below was the Pacific, usually shrouded in sea mist when we came to set up our model in the morning, but already a clear and brilliant blue speckled with white sails by the time we were ready to launch.

I often wondered what the Saturday-afternoon sailors thought of us when they saw some weird-looking object rising briefly from the test stand and blowing itself into a thousand pieces. I still keep in my desk drawer a bag of aluminum splinters which I collected after one of our flight tests, to prove to myself that all these happy memories are not just dreams.

The last and most successful of our flights took place on November 12, 1959. This was a few weeks after I had left the project and returned to my respectable scientific work in Princeton. Brian Dunne, the man who did most of our gas jet experiments, reported the event to me by letter:

Wish you could have been with us to enjoy the Point Loma festivities last Saturday. The Hot Rod flew and flew and FLEW! We don't know how high yet. Ted, who was up on the side of the mountain, guessed about 100 meters by eyeball triangulation. Six charges went off with unprecedented roar and precision. . . . The chute popped exactly on the summit and it floated down unscathed right in front of the blockhouse. . . . We are planning a champagne party for Wednesday.

So ended the second romantic age of space travel. In summer 1959 the decision was made not to use nuclear propulsion for the civilian space program, and our project was turned over to the Air Force. Ted Taylor continued his work under these military auspices, as Wernher von Braun had done in 1933. The Air Force at once put a stop to our tests of flying models. They kept the project alive for six more years, during which a great deal of good technical work was done, but the spirit and shine had gone out of it. I was at General Atomic again on the day in spring 1965 when Project Orion officially ended. We drank no champagne. The Hot Rod slept in an Air Force warehouse in Albuquerque for eighteen years and is now to be seen, looking not a day older than it did in 1959, at the National Air and Space Museum in Washington.

The U.S. Air Force did not make the mistake that Hitler made with the V-2 rocket. The Air Force tried for six years to convert an interplanetary propulsion system into a military weapon. In the end they discovered, as we had known from the beginning, that no reasonable military application of the Orion system exists. Having reached this conclusion, instead of going into mass production as Hitler did, they wisely brought the project to an end. On the day it ended I wrote a nostalgic letter to Robert Oppenheimer:

You will perhaps recognize the mixture of technical wisdom and political innocence with which we came to San Diego in 1958 as similar to the Los Alamos of 1943. You had to learn political wisdom by success, and we by failure. Often I do not know whether to be glad or sorry that we escaped the responsibilities of succeeding.

The fifteen months that I spent working on Orion were the most exciting and in many ways the happiest of my scientific life. I particularly enjoyed being immersed in the ethos of engineering, which is very different from the ethos of science. A good scientist is a person with original ideas. A good engineer is a person who makes a design that works with as few original ideas as possible. There are no prima donnas in engineering. In Project Orion, as in the safe reactor group in the little red schoolhouse, nobody was working for personal glory. It did not matter who invented what. The only thing that mattered was that the final product of our inventions should function reliably. It was a new experience for me to be caught up in a collective effort, working with a group of engineers whose whole professional life is based upon teamwork rather than on personal competitiveness. As I went happily each day to the laboratory or to the test stand on Point Loma, I remembered my mother's story of Faust among the Dutch villagers digging at the dike.

What would have happened to us if the government had given full support to us in 1959, as it did to a similar bunch of amateurs at Los Alamos in 1943? Would we have achieved by now a cheap and rapid transportation system extending all over the solar system? Or are we lucky to have been left with our dreams intact?

Sometimes I am asked by friends who shared the joys and sorrows of Orion whether I would revive the project if by some miracle the necessary funds were suddenly to become available. My answer is an emphatic no. The Test Ban Treaty of 1963, prohibiting nuclear explosions in the atmosphere and in space, made Orion flights illegal. Before one could revive Orion one would have to abrogate or renegotiate the treaty. Even without the treaty, I would not now wish to fly about in a ship that dumps radioactive debris upon the heads of the passengers in our other spaceship, Spaceship Earth. It was possible for us in 1958 to enjoy the thought of leaping into the sky with a trail of nuclear fireballs glowing behind us, because at that time the United States and the Soviet Union were testing bombs in the atmosphere at a rate of many megatons per year. We calculated that even our most ambitious program of Orion flights would add only about one percent to the contamination of the environment that the bomb tests were then causing. One percent did not seem so bad. But when I studied carefully the literature concerning the biological effects of radiation and arrived at estimates that the fallout from each Orion takeoff would statistically cause between one-tenth and one human death by radiation-induced cancer, my enthusiasm for adding even one percent to the current rate of fallout rapidly cooled. In the later years of the project, takeoff from the ground was no longer regarded as acceptable. The ship was redesigned, so that it would be carried into orbit by one or two of Von Braun's Saturn 5 rockets, and would begin exploding bombs only when it was out of the earth's atmosphere. This made the ship much more expensive and did not really solve the fallout problem. By its very nature, the Orion ship is a filthy creature and leaves its radioactive mess behind it wherever it goes. In the twenty years that have passed since Orion began, there has been a fundamental change in public standards concerning the pollution of the environment. Many things that were acceptable in 1958 are no longer acceptable today. My own standards have changed too. History has passed Orion by. There will be no going back.

The history of the exploration of space since 1958 has been the history of the professionals with their chemical rockets. The professionals have never been willing to give a fair chance to radically new ideas. Orion is dead and I bear them no grudge for that. Orion was given a fair chance and failed. But there have been several other radical schemes that came later, schemes better than Orion, schemes that could do everything Orion could do and more, schemes that do not spread radioactive debris around the solar system. None of these newer schemes has been given the chance that was given to Orion, to prove itself in fair competition with chemical rockets. Never since 1959 have the inventors of new kinds of spaceship been encouraged to try out their ideas with flying models as we did at Point Loma. You will not find any of their models resting beside our Hot Rod in the National Air and Space Museum.

The most beautiful of the unorthodox methods of space travel is solar sailing. In principle it is possible to sail around the solar system using no engine at all. All you need is a huge gossamer-thin sail made of aluminum-coated plastic film. You can trim and tack wherever you want to go, balancing the pressure of sunlight on the sail against the force of the sun's gravity to steer a course, in the same way as the skipper of an earthly sailboat balances the pressure of the wind in his sails against the pressure of the water on his keel. The idea of solar sailing has a long history. It was first imagined by the Russian pioneer of space travel, Konstantin Tsiolkovsky. It has been reinvented many times since. The latest and most elegant design for a solar sailboat is the heliogyro invented by Richard MacNeal. MacNeal's sail is a twelve-pointed star rotating like the rotor of an autogiro airplane. In 1976 the Jet Propulsion Laboratory in California made a serious attempt in cooperation with MacNeal to design an unmanned heliogyro ship that could be launched and flown in time to make a rendezvous with Halley's Comet when the comet comes by the earth in March 1986. Halley's Comet comes by only once every seventy-six years, and there is no possibility of achieving a rendezvous with chemical rockets. This was a unique opportunity for the solar sail to prove itself. The space program managers rejected the Halley's Comet mission as too risky. They cannot afford to take chances. The political consequences of a failed mission might be disastrous to their whole program. Consequently, they can never afford to support a serious exploration of radically new and untried technology. Their verdict on the solar sail proposal was rendered in the leaden prose of managerial bureaucracy:

The principal limitation preventing the Sail from receiving a positive recommendation from JPL management was the high risk associated with asserting its near term readiness in the face of absolutely no proof-of-concept tests.

When will the third romantic age in the history of space flight begin? The third romantic age will see little model sailboats spreading their wings to the sun in space, as free and graceful as the little radio-controlled gliders which dance among the birds in the sea breeze over the cliffs near the General Atomic Laboratories every Sunday afternoon. It will see test stands as amateurish as those of Berlin and Point Loma, where a new generation of young people will try out a new generation of wild ideas.

There are three reasons why, quite apart from scientific considerations, mankind needs to travel in space. The first reason is garbage disposal; we need to transfer industrial processes into space so that the earth may remain a green and pleasant place for our grandchildren to live in. The second reason is to escape material impoverishment; the resources of this planet are finite, and we shall not forgo forever the abundance of solar energy and minerals and living space that are spread out all around us. The third reason is our spiritual need for an open frontier. The ultimate purpose of space travel is to bring to humanity, not only scientific discoveries and an occasional spectacular show on television, but a real expansion of our spirit.

But space travel can only benefit the mass of mankind if it is cheap and generally available. We have a long way to go. Huge and politically oriented programs like Apollo are perhaps not even going in the right direction. I am happy to celebrate the courage of our astronauts, Gagarin and Armstrong and Aldrin and Collins and the others who came after them. But I believe the road that will take mankind to the stars is a lonelier road, the road of Tsiolkovsky, of Orville and Wilbur Wright, of Robert Goddard and the men of the VfR, men whose visions no governmental project could encompass. I am proud that I have once briefly belonged to their company.

11

Pilgrims, Saints and Spacemen

Governor William Bradford of the Plymouth Colony, President Brigham Young of the Church of Jesus Christ of Latter-day Saints, and my friend Professor Gerard O'Neill of the Princeton University physics department have much in common. Each of the three is a man of vision. Each believes passionately in the ability of ordinary men and women to go out into the wilderness and build there a society better than the one they left behind. Each has written a book to record for posterity his vision and his struggles. Each has his feet firmly on the ground in the real world of politics and finance. Each is acutely aware of the importance of dollars and cents, or pounds and shillings, in making his dreams come true.

The histories of Bradford and Young were not printed during their lifetimes but were left in manuscript form for the guidance of their followers. Bradford's manuscript was published two centuries later under the title *History of Plymouth Plantation.* Young's manuscript is quoted extensively, but not in full, in the official history of the Mormon church. O'Neill's book, *The High Frontier,* fortunately did not have to wait for posthumous publication.

The human and economic problems that the space colonists of tomorrow will face are not essentially different from the problems faced by Bradford in 1620 and by Young in 1847. Unfortunately, the extravagant style and exorbitant costs of the Apollo expeditions to the moon have created in the minds of the public the impression that any human activities in space must necessarily cost tens of billions of dollars. I believe this impression to be fundamentally mistaken. If we reject the style of Apollo and follow the style of the *Mayflower* and the Mormons, we shall find the costs of space colonization coming down to a reasonable level. By a reasonable level of costs I mean a sum of money comparable to the sums which the Pilgrims and the Mormons successfully raised.

Bradford and Young provide abundant documentation of the difficulties they faced in raising funds. Bradford emphasizes in his book that the toughest problem in the whole venture of colonization was to define a set of objectives upon which the brethren could agree:

> But as in all businesses the acting part is most difficult, especially where the work of many agents must concur, so was it found in this. For some of those that should have gone in England fell off and would not go; other merchants and friends that had offered to adventure their moneys withdrew and pretended many excuses; some disliking they went not to Guiana; others again would adventure nothing except they went to Virginia. Some again (and those that were most relied on) fell in utter dislike of Virginia and would do nothing if they went thither.

Without agreement upon objectives, the task of fund raising becomes impossible. This is a fact of life which remains as true today as it was in 1620. Bradford and Young devote more pages of their histories to the preliminary battles over objectives and finance than they devote to the description of their voyages. For both of them, it came as a blessed relief when the miseries of indecision were over, the expeditions were ready to go, and they were finally able to turn their attention away from political and financial matters to the simpler problems of physical survival. Here is Young writing from his winter quarters in February 1847, six weeks before starting his journey across the plains:

> I feel like a father with a great family of children around me, in a winter storm, and I am looking with calmness, confidence and patience, for the clouds to break and the sun to shine, so that I can run out and plant and sow and gather in the corn and wheat and say, children, come home, winter is approaching again and I have homes and wood and flour and meal and meat and potatoes and squashes and onions and cabbages and all things in abundance, and I am ready to kill the fatted calf and make a joyful feast to all who will come and partake. We have done all we could here and are satisfied it will be all right in the end.

But I must come back from these idyllic sentiments to questions of dollars and cents. Two years earlier, Young reported:

For an outfit that every family of five persons would require: one good wagon, three yoke of cattle, two cows, two beef cattle, three sheep, one thousand pounds of flour, twenty pounds of sugar, one rifle and ammunition, a tent and tent-poles—the cost would be about $250 provided the family had nothing to begin with, only bedding and cooking utensils, and the weight would be about twenty-seven hundred [pounds] including the family.

The arts were also included in Young's budget. On November 1, 1845, he paid $150 to purchase instruments for the brass band. This was a wise investment, for the band

Was sometimes invited to give concerts at villages near to the line of march, which did much to change the feelings of hostility which occasionally was manifested in such places. Thus this band proved a very great benefit to the marching column, besides cheering the spirit of the pilgrims.

The actual numbers that crossed the plains with Young were: 1,891 souls, 623 wagons, 131 horses, 44 mules, 2,012 oxen, 983 cows, 334 loose cattle, 654 sheep, 237 pigs, 904 chickens.

So we can estimate the total payload of Young's expedition to be 3,500 tons, mainly consisting of animals on the hoof, and the total cost to be $150,000 in 1847 dollars.

Bradford unfortunately does not provide such an exact accounting for the *Mayflower*. He quotes a letter from Robert Cushman, dated June 10, 1620, in London, two months before the sailing. Cushman was one of the people in charge of provisioning for the voyage:

Loving Friend, I have received from you some letters, full of affection and complaints, and what it is you would have of me I know not; for your crying out, "Negligence, negligence, negligence," I marvel why so negligent a man was used in the business.—Counting upon 150 persons, there cannot be found above £1200 and odd moneys of all the ventures you can reckon, besides some cloth, stockings and shoes which are not counted, so we shall come short at least £300 or £400. I would have had something shortened at first of beer and other provisions, in hope of other adventures; and now we could, both in Amsterdam and Kent, have beer enough to serve our turn, but now we cannot accept it without prejudice—£500 you say will serve; for the rest which here and in Holland is to be used, we may go scratch for it. —Think the best of all and bear with patience what is wanting, and the Lord guide us all.

Your loving friend, Robert Cushman

This letter shows that Cushman was personally responsible for meeting expenses to the tune of £1500. It does not say whether all the expenses, and in particular the rental fee for the *Mayflower*, were included in this figure.

Three weeks later, on July 1, 1620, an agreement was signed between the Planters and the Adventurers. The Planters were the colonists. The Adventurers were the shareholders who invested money in the enterprise and stayed at home. The agreement stipulated "that at the end of the seven years, the capital and profits, viz. the houses, lands, goods and chattels, be divided equally betwixt the Adventurers and Planters." Another clause of the agreement gave one share to each of the Planters as a bonus for their seven years of hard labor: "Every person that goeth being aged 16 years and upward be rated at £10, and £10 to be accounted a single share." Any cash that the Planters contributed would entitle them to additional shares.

The 1620 agreement proved unsatisfactory to both sides and caused constant friction. In 1626, a year before the planned division of assets, the whole matter was renegotiated and a new agreement was signed, "drawn by the best counsel of law they could get, to make it firm." The 1626 agreement stipulated that the Adventurers sell to the Planters, "in consideration of the sum of one thousand and eight hundred pounds sterling to be paid in manner and form following, —all and every the stocks, shares, lands, merchandise and chattels— any way accruing or belonging to the generality of the said Adventurers aforesaid." Having bought out the Adventurers' shares, the Planters were left with a debt of £1800, which they finally succeeded in paying off twenty-two years later.

I do not know how much profit or loss the Adventurers took in the 1626 settlement. I also do not know how large a fraction of the original cost of the expedition was paid by the Planters. As to the first point, it is unlikely that the Adventurers took a loss, for the colony was not bankrupt in 1626 and the Adventurers were not in the habit of lending their money for nothing. As to the second point, it is unlikely that the Planters paid as much as half of the original costs. If they had been in a position to pay half, they would probably have managed to squeeze the expenses down to such a point that they could do without the Adventurers altogether and avoid the innumerable headaches that the partnership brought with it. I therefore conclude from the evidence of the 1626 settlement that £3600 is a safe upper limit to the original cost of renting and provisioning the *Mayflower*. The evidence of the Cushman letter implies a lower limit of £1500. I shall adopt £2500 as my estimate of the cost of the expedition in 1620 pounds. This figure can hardly be wrong by a factor of two either way. The payload of the *Mayflower* is stated explicitly by Bradford. It was 180 tons.

My next problem is to convert the 1620 and 1847 cost figures into their modern equivalents. A good source of information about the history of wages and prices in England is the work of Ernest Phelps Brown and Sheila Hopkins, published in two articles in the journal *Economica* and reprinted in a series called *Essays in Economic History*, put out by the Economic History Society. The first article deals with wages, the second with prices. It is a question of taste whether one prefers to use wages or prices as the basis for comparing costs between different centuries. If we use wages, we are saying that an hour of a workingman's time in 1620 is equivalent to an hour in 1979. If we use prices, we are saying that a pound of butter in 1620 is equivalent to a pound of butter today. My personal opinion is that wages give a truer standard of comparison than prices. My purpose in making the comparison is to try to define in a roughly quantitative fashion the size of the human efforts that the *Mayflower* and the Mormon expeditions demanded.

According to Phelps Brown and Hopkins, the wages of workers in the building trade in 1620 were in the range from 8 to 12 pence per day. In 1847 the range was from 33 to 49 pence. For the modern equivalent of these numbers I take the minimum rate of $9.63 per hour imposed by building trade union contracts in New York in 1975. The exchange rates on the basis of wages are then:

$$£1 \text{ (1620) equals } \$2500 \text{ (1975)}$$
$$\$1 \text{ (1847) equals } \$100 \quad \text{(1975)}$$

These are very approximate numbers. A rough check on the numbers for 1620 is provided by the fact, already mentioned, that each Planter received a credit of £10 for going to Plymouth and working for the community for seven years without wages.

The estimated total costs in 1975 dollars are then 6 million for the *Mayflower* and 15 million for the Mormons. On this basis I have drawn up the first two columns of Table I. The point I am trying to emphasize with these numbers is that both the *Mayflower* and Mormon expeditions were extremely expensive operations. In their time, each of them stretched the limits of what a group of private people without governmental support could accomplish.

The numbers in the bottom row of Table I give an estimate of the number of years an average wage earner would have had to save his entire income to pay the passage for his family. Although the average Mormon family was twice as large as the average *Mayflower* family, the cost in man-years per family was three times as large for the *Mayflower* as it was for the Mormons. This difference had a decisive effect on the financing of the colonies. An average person, with single-minded dedication to a cause and with a little help from his friends, can save two or three times his annual income. An average person with a family to feed, no matter how dedicated he may be, cannot save seven times his income. So the Mormons were able to pay their way, while the Planters on the *Mayflower* were forced to borrow heavily from the Adventurers and to run up debts which took twenty-two years to pay off. Somewhere between two and seven man-years per family comes the breaking point, beyond which simple do-it-yourself financing by ordinary people becomes impossible.

TABLE I Comparison of Four Expeditions

(Cost exchange rates based on building trade wages)

(M means millions)

Expedition	Mayflower	Mormons	Island One L5 Colony	Homesteading the Asteroids
Date	1620	1847	1990 +	2000 +
Number of People	103	1891	10000	23
Payload (tons)	180	3500	3.6M	50
Payload (tons) per person	1.8	2	360	2
Cost (1975 dollars)	$6M	$15M	$96000M	$1M
Cost per pound (1975 dollars)	$15	$2	$13	$10
Cost in man-years per family	7.5	2.5	1500	6

I said nothing yet about the last two columns in my table. These represent two contrasting styles of space colonization, both taken from O'Neill's book, with some changes for which I am responsible. Column 3 comes from O'Neill's Chapter 8, which he entitles "The First New World," describing space colonization organized by the American government in the official NASA style. Column 4 comes from O'Neill's Chapter 11, with the title "Homesteading the Asteroids," in which he describes space colonization done in the *Mayflower* style by a bunch of enthusiastic amateurs.

The cost of the "Island One" project is $96 billion. Many people, myself included, feel that $96 billion is a preposterously large amount of money to spend on any single enterprise. But still we have to take this number seriously. It was arrived at by a group of competent engineers and accountants familiar with the ways of the government and the aerospace industry. It is probably the most accurate of all the cost estimates that I have included in Table I. For this $96 billion you can buy a great deal of hardware. You can buy a complete floating city to house and support ten thousand people with all modern conveniences at the magic point L5, which is just as far from the earth and from the moon as these bodies are from each other. You can buy enough synthetic farmland to make a closed ecological system which supplies the colonists with food and water and air. You can buy a spaceborne factory in which the colonists manufacture solar power stations to transmit huge amounts of energy in the form of microwave beams to receivers on the earth. All these things may one day come to pass. It may well be true, as O'Neill claims, that the investment of $96 billion will be repaid in twenty-four years out of the profits accruing from the sale of electricity. If the debt could be paid off in twenty-four years, that would be almost as quick as the *Mayflower* Planters could do it. But there is one inescapable difference between Island One and the *Mayflower.* The bottom row of Table I shows that the Island One colonist would have to work for 1500 years to pay his family's share of the costs. This means that Island One cannot by any stretch of the imagination be considered as a private adventure. It must inevitably be a government project, with bureaucratic management, with national prestige at stake, and with occupational health and safety regulations rigidly enforced. As soon as our government takes responsibility for such a project, any serious risk of failure or of loss of life becomes politically unacceptable. The costs of Island One become high for the same reason that the costs of the Apollo expeditions were high. The government can afford to waste money but it cannot afford to be responsible for a disaster.

After this brief visit to the superhygienic welfare state at Island One, let us go on to the last column of Table I. The last column describes O'Neill's vision of a group of young pioneers who save enough money to move out on their own from the L5 colony into the wilderness of the asteroid belt. They are going on a one-way trip at their own risk. The cost estimates here describe hopes rather than facts. Nobody can possibly know today whether it will be feasible for a group of twenty-three private people to equip such an expedition at a total cost of a million dollars. Anybody who is professionally qualified to estimate costs will say that this figure is absurdly low. I do not believe that it is absurdly low. It is no accident that the per capita cost estimates for the asteroid colony turn out to be similar to those of the *Mayflower*. This is the maximum level of costs at which the space beyond the earth will give back to mankind the open frontier that we no longer possess on this planet.

According to the third and fourth columns of Table I, the cost per pound of the asteroid expedition is not significantly less than that of Island One. The big differences between the two expeditions lie in the number of people and in the weight carried per person. The feasibility of cheap space colonization in the style of the asteroid expedition depends upon one crucial question. Can a family, bringing a total weight of only two tons per person, arrive at an asteroid, build themselves a home and a greenhouse, plant seeds and raise crops in the soil as they find it, and survive? This is what the *Mayflower* and Mormon colonists did, and it is what the space colonists must do if they are to be truly free and independent.

No space probe has yet visited an asteroid. No scientific instruments have even been flown by an asteroid to give us a closer look at it. We are still as ignorant of the topography and chemistry of the asteroids as we were ignorant of the topography of Mars before the Mariner and Viking missions. Until some of the asteroids have been surveyed with unmanned instruments, it is pointless to try to foresee in detail the problems that colonists would face in making themselves at home there. Cost estimates for farming on an asteroid are meaningless until we know whether the soil is soft enough to be dug without using dynamite. Instead of speculating about the mechanics of space colonization in an unknown environment, I will only mention some institutional reasons why it may not be absurd to imagine a reduction in costs by a factor of 100,000 from the $96 billion of Island One to the $1 million of the asteroid colony. First we save a factor of four hundred by reducing the number of people from ten thousand to twenty-three. That leaves a factor of 250 still to be found. We may hope to save a factor of ten by accepting risks and hardships that no government would impose upon its employees, and another factor of five by eliminating trade union rules and bureaucratic management. The last factor of five will be harder to find. It might come from new technology, or more probably from salvaging and reusing equipment left over from earlier government projects. There are already today several hundred derelict spacecraft in orbit around the earth, besides a number on the moon, waiting for our asteroid pioneers to collect and refurbish them.

The Island One and the asteroid homesteading expeditions are extreme cases. I chose them to illustrate high and low estimates of the costs of colonization. The true costs, when colonization begins, will probably lie somewhere in between. In so difficult and long-range a venture, there is room for a mixture of styles. Governmental, industrial and private operations must all go forward, learning and borrowing from one another, before we shall find out how to establish colonies safely and cheaply. The private adventurers will need all the help they can get from governmental and commercial experience. In this connection, it is worth remembering that 128 years passed between the voyages of Columbus and the *Mayflower*. In those 128 years, the kings and queens and princes of Spain and Portugal, England and Holland, were building the ships and establishing the commercial infrastructure that would make the *Mayflower* possible.

O'Neill and I have a dream, that one day there will be a free expansion of small groups of private citizens all over the solar system and beyond. Perhaps it is an idle dream. It is a question of dollars and cents, as Bradford and Young well knew. We shall never find out what is possible until we try it.

12

Peacemaking

During the year that I worked for the Orion project, a great debate was raging over the question of bomb testing. Should we, or should we not, try to negotiate with the Soviet Union an agreement to stop nuclear testing completely? My old friend Hans Bethe was pushing hard, in public and within the government, for a comprehensive test ban. My new friend Edward Teller was pushing hard against a ban. My affection and respect for Bethe never wavered, but in this debate I was heart and soul on Teller's side. Orion could not survive without bomb tests. In the short run, we needed at least one test to convince the skeptics that our ship could take the blast from a nuclear explosion a hundred feet away and remain intact. In the long run, we needed many more tests to develop fission-free bombs, so that the radioactive fallout from our voyages would be reduced almost to zero. I knew that my own objectives in working for Orion were peaceful and pure, and I did not see any justice in labeling Teller a warmonger merely because of his passionate desire to explore to the end the thermonuclear technology that he had pioneered. Teller and I fought together with a good conscience against the test ban. I was sorry to see the good Hans Bethe fighting on the wrong side. I worried over the possibility that the security people would punish him for his errors of judgment in this matter, as they had punished Oppenheimer five years earlier.

In the summer of 1959, as my time with Orion was coming to an end, I tried to do what I could to improve the project's chances of survival. I made a pilgrimage with Ted Taylor to Jackass Flat, the desert area in Nevada where we hoped to carry out our first crucial demonstration of feasibility with a real bomb. I went for two weeks to Teller's weapons laboratory at Livermore and worked there with the team that was trying to design fission-free weapons. And I wrote an article for publication in the respected political journal *Foreign Affairs*, arguing against the test ban with all the eloquence I could muster.

Only once in my life have I experienced absolute silence. That was Jackass Flat under the midday sun. Long ago I read in Herbert Ponting's *The Great White South* of the silence of a windless day in Antarctica. Jackass Flat was as silent as Antarctica. It is a soul-shattering silence. You hold your breath and hear absolutely nothing. No rustling of leaves in the wind, no rumbling of distant traffic, no chatter of birds or insects or children. You are alone with God in that silence. There in the white flat silence I began for the first time to feel a slight sense of shame for what we were proposing to do. Did we really intend to invade this silence with our trucks and bulldozers, and after a few years leave it a radioactive junkyard? The first shadow of a doubt about the rightness of Orion came into my mind with that silence.

Nevertheless, I went a few weeks later to Livermore, with ambitious plans to explore the possibilities of fission-free bombs. For two weeks I worked hard, trying to design a bomb that would reduce the fallout from Orion by a factor of ten. This was the only time in my life that I worked directly as a bomb designer. I was there only because I wanted to explore the universe, and there was no thought of murder in my heart. But I learned at Livermore that it is not possible to make a clean separation between peaceful and warlike bombs, or between peaceful and warlike motives. Motives in each of us tend to get mixed. The colleagues with whom I worked at Livermore were inventing devices that later came to be known as neutron bombs. I helped them and they helped me. In two weeks I made friends with them and became to some extent one of their team. To that extent I share the responsibility for the existence of neutron bombs. After this experience I could never again honestly say that the bombs we wanted to use for Orion had nothing to do with bombs that are designed for killing people.

The *Foreign Affairs* article was called "The Future Development of Nuclear Weapons" and was accepted enthusiastically by the editors. It appeared in April 1960. The main thesis of the article is that a permanent test ban would be a dangerous illusion because future improvements in weapons technology would create irresistible pressures toward secret or open violation of any such ban. In other words, fission-free bombs are the wave of the future, and any political arrangement which ignores or denies their birthright is doomed to failure. Here is a fair sample of the rhetoric to which the editors of *Foreign Affairs* gave their approval:

> Imagine a hypothetical situation in which the United States is armed with its existing weapons, while some adversary (not necessarily the Soviet Union) has a comparable supply of nuclear fuel and has learned how to ignite it fission-free. The adversary's bombs would then outnumber ours ten or a hundred to one, and theirs could be used with far greater versatility in infantry warfare. . . . Any country which renounces for itself the development of nuclear weapons, without certain knowledge that its adversaries have done the same, is likely to find itself in the position of the Polish army in 1939, fighting tanks with horses.

I cannot excuse this effusion on the grounds that it was written as a last desperate attempt to save Orion from extinction. Obviously there was more to it than that. It was written, in so far as I can be aware of my own motives, as an act of personal loyalty to Edward Teller and to his colleagues with whom I worked at Livermore. I was deeply impressed by the fragility of the efforts at Livermore to design radically cleaner explosives. Inside the barbed-wire fence at Livermore, all the serious thinking was being done by five or six gifted young people, who worked under depressing conditions of physical and mental isolation. They might at any moment decide to quit. Outside the fence, the whole society was indifferent or actively hostile to their efforts. My article was in some sense an act of psychological atonement which I owed to Edward Teller for the fact that I was leaving him and going back to Oppenheimer at Princeton. I wanted to show my friends at Livermore that there was at least one person outside the fence who cared.

In retrospect it is easy to see that my argument was wrong on at least four counts: wrong technically, wrong militarily, wrong politically and wrong morally. Technically, I misjudged the time scale for development of fission-free weapons. I expected they would be generally available within ten years. More than ten years have passed by without any visible sign of them. Militarily, I was wrong in thinking of "tactical nuclear war" as a feasible way to employ military forces. Since 1960 I have taken part in some detailed studies of tactical nuclear war, and I have seen the results of some war games played by professionals. The evidence convinces me that a tactical nuclear war conducted between any two nuclear powers will quickly degenerate into an uncontrollable chaos that can be ended only by an immediate cease-fire (if we are lucky) or by an escalation to strategic strikes (if we are unlucky). In either case the outcome of the war will hardly be affected by the presence of fission-free weapons on one or both sides of the initial conflict.

Politically, I was wrong in saying that a test ban would surely be ineffective as a means of stopping development of fission-free weapons. A total test ban would at least stop our side from developing these weapons. If it were known that we had stopped work on them and did not consider them to be militarily important, the incentive for the other side to put serious effort into developing them would be greatly reduced. On the other hand, the one way to make certain that our adversaries would soon possess these weapons would be for us to develop and deploy them ourselves.

Morally, I was wrong in accepting unquestioned the morality of supplying our own soldiers with new weapons. Vietnam has taught us that our weapons are not always wisely used. In spite of all our mistakes in Vietnam, we avoided the supreme mistake of using nuclear weapons there. If our soldiers in Vietnam had been supplied with small fission-free nuclear weapons, the pressure to allow use of these weapons at moments of crisis would have been hard to resist. The consequence might easily have been a tragedy in Vietnam far greater even than the tragedy we have witnessed.

It seems obvious now that the *Foreign Affairs* article was a desperate attempt to salvage an untenable position with spurious emotional claptrap. Yet the surprising fact is that before I submitted it to *Foreign Affairs* I showed it to two of the wisest men I knew, Robert Oppenheimer and George Kennan, asking for their comments. George Kennan, after a distinguished career as a diplomat, had become a historian and a colleague of mine at the Institute for Advanced Study in Princeton. Both Oppenheimer and Kennan read the article and encouraged me to publish it. Perhaps, after all, even the best of us are a little wiser now than we were in 1960.

Oppenheimer changed his mind about the article rather soon

after it appeared in print. He wrote to me, as usual, cryptically, quoting a Hungarian proverb: "It is not enough to be impolite; one must also be wrong."

By that time I had finally become an American citizen. The decision to abjure my allegiance to Queen Elizabeth might have been a difficult one, but the Queen's ministers made it easy for me. An official lady in the Queen's Foreign Office decided that my children were illegitimate according to British law. They were therefore not British subjects and not entitled to receive British passports. As a consequence of her decision, my family for a while consisted of five people with five different nationalities, one British, one German, one Swiss, one American and one stateless. Traveling around Europe with a stateless child is no joke. So it was with considerable relief that I went to the courthouse in Trenton and said the magic words that released me from dependence on any foreign prince or potentate. Bastards or not, the U.S.A. would at least give my children passports.

As a newborn American I was quick to exercise the privileges of citizenship and became active in the Federation of American Scientists, a political organization which lobbies in Washington for various good causes. The federation had a Washington office run by Daniel Singer, a lawyer retained by the federation with the title of general counsel. Singer was doing part time the job that is now being done full time by Jeremy Stone. In 1960 I was elected to the council of the federation, and received from Singer an education in the fine points of congressional politics. Singer welcomed the fact that my *Foreign Affairs* article had given me a reputation as a military hard-liner. He said the federation's main problem was that its spokesmen were usually such notorious liberals that their opinions were discounted in advance.

In 1961 the federation was trying to help push through Congress the bill establishing a new department of the U.S. government, the Arms Control and Disarmament Agency. Kennedy had intended to set up ACDA as soon as he became President, believing that this would help him to conduct arms control and disarmament negotiations in a more professional and less haphazard style than we had usually followed in the past. But he had difficulty getting ACDA approved by Congress. On the last day before Congress was due to adjourn for the September recess, the ACDA bill had still not passed the Senate and it seemed likely that it would not even come up for a vote. In desperation Singer looked through the list of federation members for a name that the conservative wing of the Senate might listen to. He found the name of Herman Kahn, whose book *On Thermonuclear War* had recently appeared and made his reputation as a military hard-liner even more secure than mine. Singer telephoned Kahn and asked him if he would come to Washington immediately to save the ACDA bill. Kahn, being himself a professional arms controller, thought that ACDA was needed. He came at the last moment before the Senate committee and argued for ACDA in language that the most conservative Senators could understand. The bill passed, and the Senators ran for their planes.

ACDA was hastily organized at the beginning of 1962. The head of the Science and Technology Bureau was Frank Long, a chemist who had been recruited from Cornell University. Long had somehow to collect within a few months a staff of competent scientists. He decided it would be a good idea to offer some temporary summer appointments in his bureau. The people who came for the summer would do no great harm if they were incompetent and might be persuaded to stay if they were competent. Dan Singer asked me if I would like to apply for one of these summer jobs. I had an interview with Long and was accepted. So in June I went to work at ACDA. I worked there for two summers, 1962 and 1963. After 1963 the agency had an adequate permanent staff and migrant workers were no longer needed.

ACDA in 1962 and 1963 was a delightful place to work. The agency had the status of a department of government, but contained altogether only about fifty people. In the Science and Technology Bureau there were only ten of us. We had not yet had time to become bureaucratic. We sat in big old-fashioned offices on the ground floor of the old State Department building. Every morning, copies of the diplomatic telegrams of the previous twenty-four hours were circulated for us to read. Sometimes I felt a little nervous, seeing the telegrams lying on somebody's desk under a window, within easy view of the pedestrians walking along the street outside. The building dates from the spacious old times when Henry Stimson as Secretary of State opposed the creation of an American cryptological office for breaking foreign codes. "Gentlemen do not read each other's mail," said Stimson. Measured by this standard, my colleagues and I were not gentlemen. We enjoyed each morning the latest gossip about the Soviet party secretary with matrimonial problems or the important diplomat's daughter found dead drunk in the streets of Paris. A few of the telegrams were more serious and discussed details of negotiations in progress.

The main business of ACDA in summer 1962 was to prepare positions for two sets of negotiations, the test-ban negotiations with the Soviet Union and the disarmament negotiations at the Eighteen-Nation Disarmament Conference organized by the United Nations. The old hands all knew that the test-ban negotiations were for real while the disarmament negotiations were nothing but an exercise in propaganda. Most of the young recruits wanted to be involved with the immediate problems of the test ban. Frank Long knew that I was unenthusiastic about the test ban, and so he suggested that I should spend my two months studying longer-range problems of disarmament. He wanted me to see whether there might not after all be some opportunities for the American delegation to push the Eighteen-Nation Conference into doing something useful.

The main problem with the Eighteen-Nation Conference was that the Soviet delegation talked about "general and complete disarmament" while the Western delegations talked about limited and specific reductions of forces. In order to seem responsive to the Soviet proposals, the United States had offered its own plan for general and complete disarmament by stages. In our plan, Stage One had to be satisfactorily completed before we would be committed to Stage Two. Everybody knew that Stages Two and Three were pure moonshine. It would be a major miracle if we ever got to the end of Stage One.

The one person at ACDA who took general and complete disarmament seriously was Louis Sohn, a Harvard lawyer who specialized in international law. I spoke frequently with Sohn and learned a great deal. He was promoting a scheme called "zonal disarmament," which he offered as a fair compromise between Soviet and Western positions. The rules of zonal disarmament were as follows: Each country should divide its own territory into a certain number of zones. At the beginning of each year one zone should be opened to a force of international inspectors, who would supervise the dismantling and destruction of all weapons found in the zone. The choice of the zone to be opened should be made either by an adversary country or by a random process. Thus it would be to the advantage of each country to distribute its weapons evenly among its zones. There were in addition various special rules and exemptions to deal with capital cities and unique military facilities. This was the "Sohn plan," which was in those days popular in liberal intellectual circles.

One of the first things I did at ACDA was to work out a variant of the Sohn plan, which I called "progressive geographical disarmament." It seemed to me that Sohn's idea of turning disarmament into a two-person game was too logical, better suited to academic game theorists than to real-world politicians. So I simplified Sohn's plan by removing the game-theoretical features. A progressive geographical disarmament treaty would require each country to divide its territory into zones of equal area. At the beginning of each year one zone would be opened to inspection and demilitarization, but the choice of the zone would be made by the owner of the territory. In this way I hoped to make the international inspection less intrusive and less objectionable to Soviet sensibilities. There would be plenty of time for each country to remove militarily sensitive or politically embarrassing material from a zone before the inspectors came to look at it. Dirty linen could be privately washed and bloodstains on the wall painted over. I discussed the details of my plan with Sohn and wrote it down in an official ACDA memorandum. Proudly I offered it to Frank Long as my solution of the disarmament problem. It disappeared into the ACDA files and was never seen again.

In 1961 and 1962 the United States and the Soviet Union exploded more bombs than ever before. Many of them were megaton bombs, and the fallout radioactivity was rising alarmingly all over the Northern Hemisphere. One quiet evening in my office at ACDA, I collected information about the tests and drew a simple diagram on graph paper to show what was happening. From left to right I plotted the years from 1945 to 1962. Above each year I plotted vertically the cumulative total number of all bombs exploded from 1945 to that year. As soon as the diagram was finished, the situation became clear. The curve of cumulative bomb totals was an almost exact exponential, all the way from 1945 to 1962, with a doubling time of three years. A simple explanation suggested itself for this doubling every three years. It takes roughly three years to plan and carry out a bomb test. Suppose that every completed bomb test raises two new questions which have to be answered by two new bomb tests three years later. Then the exponential curve is explained. Having discovered this profound truth about bomb tests, I was ready to draw the consequences. Some questions have to remain unanswered. At some point, we have to stop. That evening I accepted for the first time the inevitability of a test ban.

On the Fourth of July I went with my wife and our two youngest children to watch the fireworks on the Ellipse behind the White House. A big crowd was there, predominantly black, sitting on the grass and waiting for the show. We sat down among them. Our children were soon running around with the others. Then came the fireworks. After the official fireworks were over, the crowd was allowed to let off unofficial fireworks. Everybody seemed to have brought something. The black children all had little rockets or Catherine wheels or sparklers and were shouting with joy as they blazed away. Only our children sat quiet and sad because we had not brought anything for them. But suddenly one of the black children came up to us and gave our children a fistful of sparklers so they could join in the fun. That moment, rather than the ceremony in Trenton, was the true beginning of my citizenship. It was then that I knew for sure we were at home in America.

I spent the second half of summer 1962 at ACDA making an intensive study of Soviet policies and attitudes. Frank Long thought that a few weeks of immersion in Soviet documents would give me a more realistic view of the problems of disarmament. I had inherited from Frank Thompson an abiding love for the Russian language and an ability to read it with reasonable fluency. I found in ACDA an excellent collection of source material, Russian newspapers and military and political publications. Also, Raymond Garthoff was there, an expert Sovietologist who helped me find my way among the files. I wanted to get inside the skin of the Soviet leaders and see the world as they saw it; afterward I could perhaps provide ACDA with some useful guidance in dealing with them.

Above all else, I read every utterance of Khrushchev that I could lay my hands on. Khrushchev I found invaluable. Unlike other official Russians, he spoke from the heart. No hack speech-writer would have dared to write for him the things he said: often inconsistent, often bombastic, surprisingly often human and personal. I had a strong sense that this was a unique moment in history, when a man so open and so whimsical was in power in Russia. If we did not start quickly to negotiate with him about basic issues in a language he could understand, the opportunity might be gone forever.

My study of the Soviet literature convinced me that the Russians were totally serious about maintaining the superiority in conventional forces, infantry, tanks and guns that had brought them victory in World War II. They were serious about civil defense, organized as a major activity of the citizens' sports and training movement, DOSAAF, which was important for maintaining a feeling of solidarity between the civilian population and the armed forces. They were not, in the same sense, serious about the advanced technological weapons that dominated American thinking. Khrushchev poured immense quantities of money into the development of an antiballistic missile defense system with huge radars and long-range interceptor rockets, but he did not really care whether the thing would work. The idea of cost effectiveness, so central to our thinking, was absent from his view of the world. Our experts and politicians worried about Soviet secret weapons, arguing that if the Soviet ABM system was really as ineffective as it looked, Khrushchev would not be building it. I knew better. Khrushchev once said that he had wanted to make public a film of a test of his ABM system, but his advisers persuaded him not to do it. Khrushchev was evidently thinking of the weapon as a political showpiece while the advisers were more seriously concerned with its deficiencies as a military system.

Khrushchev's ABM system was only the latest example of a long Soviet tradition of defense by bluff, the exploitation of advanced weapons of dubious military value for political and psychological purposes. There were public displays of massed parachute landings in the 1930s, and public displays of advanced jet bombers in the 1950s. The prompt conversion of the first Soviet intercontinental missile into a booster for Sputniks was in the same tradition. In each case the Soviet Union took the opportunity provided by a new weapon to make an impressive show of strength. The weapons displayed were in fact prototypes, but the public was given the impression that they were in mass production. The Soviet leaders were able, without actually lying, to exaggerate their strength and distract attention from their weaknesses. The existence of rigid internal secrecy in the Soviet Union made such tactics possible and effective.

In the fall of 1962, between my two summers at ACDA, Khrushchev astonished the world by trying to place nuclear missiles in Cuba. According to my view of Khrushchev's character, this venture was another example of his not taking advanced weapons seriously. He probably thought of the missiles in purely political terms, as an impressive show of strength with which he could give political support to his Cuban ally. He did not realize that Kennedy would think of the missile deployment as a military move open to military countermeasures and would use military means to frustrate it. When the missile crisis was over, all right-thinking Americans believed that Kennedy's handling of it had been masterly and heroic. Even if I had known in the summer of 1962 what was to happen in October, I could never have hoped to persuade the senior officials in ACDA to accept my opinion that the missiles in Cuba were only a typical Soviet defense by bluff, which Kennedy was under no compulsion to demolish.

At the end of that summer I wrote a memorandum, summarizing what I had learned about the Soviet ABM system and recommending a sharp change in the official American response to it. In the past, I said, America reacted very stupidly to Soviet attempts at defense by bluff. We failed to understand that it is to our advantage to be facing a defense by bluff rather than a militarily real defense, even when the quality of our intelligence is not good enough to tell the difference. For example, in 1960 we enjoyed a superiority in offensive missiles while the Soviet Union concealed its weakness by maintaining a missile bluff. We then demolished the Soviet missile bluff as conspicuously as possible with public statements of the results of U-2 photography, and so forced the Soviet Union to replace its fictitious missile force by a real one. It would have been much wiser for us to have left the Soviet bluff intact.

For the future, I argued that the United States should strive by every means in its power to sustain and buttress the Soviet ABM bluff. We should try to discourage the American Secretary of Defense from making loud public statements of the Soviet system's ineffectiveness. We should not contest Khrushchev's claims of technological superiority in this area. In our negotiations with the Soviet Union we should seek limitations only on offensive weapons which would be threatening to us, and should avoid any prohibition of deployment of ABM systems. If, following our past pattern of behavior, we were to talk the Soviet leaders into abandoning their ABM system, we would be forcing them to transfer vast technological resources from a harmless defense by bluff into far more dangerous and militarily effective weapons systems.

I handed my second memorandum, with the title "U.S. Reaction to Soviet Ballistic Missile Defense," to Frank Long. This time I had made an accurate assessment of Soviet actions and motives. But I had completely ignored the half of the world with which Long was mainly concerned. I had forgotten American domestic politics. How could I ask Secretary of Defense McNamara to stand before Congress and praise the Soviet ABM system, when his listeners would immediately seize upon his words as a confession of the criminal negligence of the Kennedy administration in letting the Russians get ahead of us? My second memorandum, like my first, disappeared into the ACDA files.

After a year in Princeton I came back to ACDA for the summer of 1963. The atmosphere was completely changed. The final stages of the test-ban negotiations were about to begin in Moscow. ACDA was going into action, and all hands were needed to fight the impending battle. I gladly put aside the long-range analysis of Soviet strategic doctrines and joined the test-ban team. Frank Long and some of the other senior ACDA people went to Moscow to help Averell Harriman negotiate the treaty. Those who stayed behind at ACDA had the job of preparing positions for the second stage of the battle, the fight for ratification of the treaty by the United States Senate.

I had my little moment of glory shorty before the treaty was signed. One of the stickiest points in the negotiations was whether or not to include peaceful nuclear explosions in the prohibition of tests in the atmosphere. The American negotiating position was that peaceful explosions should be allowed. The Russians said no, and refused to budge. The American position was designed to win votes for ratification from senators who strongly supported Project Plowshare, a Livermore program which aimed to dig canals and harbors with nuclear explosives. The Russians claimed that Plowshare was only a subterfuge for continuing weapons tests under another name. The negotiations were deadlocked for several days. Harriman cabled back to Kennedy in Washington, "I think we shall have a treaty if I give way on this one." Kennedy, so I was told, picked up the phone and asked William C. Foster, the director of ACDA, what he thought about it. Foster said he would like to talk it over with his experts. Foster called the ACDA Science and Technology Bureau and spoke with Al Wadman, one of the bureau staff. It was late in the afternoon and almost everyone had gone home. Wadman and I were the only people left in the office. Wadman came over to me and asked me if I thought we should stand firm on peaceful explosions. I was the only person in ACDA who had been at Livermore and knew something at first hand about Plowshare. At that moment I was thinking not so much about Plowshare as about Orion. I said to Wadman, "Of course we should give way." Wadman called back Foster and Foster called back Kennedy and the cable went back to Harriman and the treaty was signed.

This story gives a misleading impression of what happened, even if it happens to be true. No doubt Kennedy called others besides Foster and Foster called others besides Wadman. I am sure that if I had given Wadman a different answer the treaty would still have been signed. The treaty had lain for a long time in the womb of history and had come to the day of its birth. We were only the midwives.

Two days later I met Ted Taylor in Washington and told him I had signed Orion's death warrant. Ted took the news calmly. He, too, had known for some time that his five-year struggle to keep Orion alive was coming to its inevitable end.

My next service to ACDA was to pay a visit to the director of the Plowshare program at the headquarters of the Atomic Energy Commission. I went with Wadman to extract from the director a written statement saying whether or not his program could continue within the terms of the treaty as signed. The director was faced with a dismal choice. If he said yes, he was helping to ratify the treaty. If he said no, and the treaty was ratified in spite of him, his program would probably be closed down. Bureaucratic politics is a dirty game, even when the good guys are winning. We had him neatly skewered, and he knew it. He said yes, and we took his statement back in triumph to ACDA.

At the end of August, the treaty ratification hearings began before the Senate Foreign Relations Committee, with Senator Fulbright as chairman. Edward Teller testified eloquently against the treaty. I was invited to testify in favor of the treaty, speaking not as an ACDA employee but as a private citizen representing the Federation of American Scientists. Daniel Singer was on friendly terms with the Fulbright committee staff and arranged the invitation. He thought that I, a defector from the enemy camp, would be more effective as a witness than one of the federation stalwarts who had been fighting for the test-ban from the beginning.

I had the good luck to speak immediately after George Meany, the president of the American Federation of Labor. The senators appeared in force to listen to Meany, and most of them stayed to listen to me. Meany was speaking for fifteen million voters, I for two thousand. Meany gave a stirring performance. He fulminated for fifteen minutes against the Russians, describing the contempt and distrust with which the honest laborers of America regarded the deceitful Communist negotiators. Then, right at the end of his speech, he came on with his punch line. But, he said, the honest laborers of America also have to think of their wives and children. They have to protect their wives and children from the poison that falls from the sky as a result of bomb tests. So for the sake of their wives and children, and in spite of their distrust and contempt for the Communists, the honest laborers of America support the treaty.

That was a hard act to follow, but I made my little speech and the senators listened attentively. I explained briefly what I had learned at ACDA and through contacts with Russian scientists about the nature of Soviet society, and I described the disastrous effect that a failure to ratify the treaty would have on the people within the Soviet establishment who believed in peaceful coexistence. When I had finished, Senator Fulbright asked me one question, knowing well what my answer would be. What precisely did Khrushchev mean when he said "We shall bury you"? I replied that in Russian this phrase is commonly used with the meaning "We shall be here to celebrate your funeral." It means simply "We shall outlive you" and does not imply any murderous intentions.

The day after my Senate testimony, I took another half day off from my job at ACDA and strolled down from the State Department building onto Constitution Avenue, a few blocks away. There another kind of history was being made. Black people from all over the United States were marching. A quarter of a million people were marching. It was quiet. No music and no stamping of feet. I walked to the end of the avenue where the marchers were assembling and marched with them to the Lincoln Memorial. Each group of people carried banners saying where they came from. Occasionally there would be cheering and shouting from the crowd when a group came by from one of the really tough places—Birmingham, Alabama, or

Albany, Georgia, or Prince Edward County, Virginia, the battle grounds of the early freedom fighters. The people from the Deep South were very young, hardly more than children. The Northerners were older, many of them husbands with their wives, or union members brought to Washington by their unions. In those days, in the Southern towns where the battles for civil rights had been raging, black people with family responsibilities could not afford to take chances. From the toughest places only young people came.

Most of these children from the Southern battle grounds had never been away from their homes before. They had been fighting lonely battles. They had never before had anybody to cheer for them. They had never known that they had so many friends. They sang their freedom songs while the Northerners listened, and they looked like the hope of the future as they danced and sang with their bright faces and sparkling eyes.

From two till four, the leaders of the black people spoke at the memorial, with the huge figure of Lincoln towering over their heads. Only James Farmer did not speak, but instead sent a message from his cell in a Louisiana jail. Martin Luther King spoke like an Old Testament prophet. I was quite close to him and I was not the only one listening who was in tears. "I have a dream," he said, over and over again, as he described to us his visions of peace and justice. In my letter to my family that night I wrote, "I would be ready to go to jail for him any time." I did not know then that I had heard one of the most famous speeches in the history of mankind. I only knew that I had heard one of the greatest. I also did not know that Martin Luther King would be dead within five years.

It would be difficult to find two human beings more different than George Meany and Martin Luther King, the old plumber from the Bronx and the young prophet from Atlanta. But their differences were not so important as what they had in common. Both were tough. Both became leaders by demanding justice for their people. Both believed in the future, and in children. Each, in his own fashion, was a peacemaker.

13

The Ethics of Defense

On the one side, the gospel of nonviolence that Jesus and Gandhi and Martin Luther King preached and practiced. On the other side, the madness of hydrogen bombs and the doctrine of Mutual Assured Destruction with which we are now precariously living. Given this choice, how could any sane person not choose the path of nonviolence? I made the choice once, when I was fifteen years old, in the days of Cosmic Unity. Then the choice seemed simple. I would die for Gandhi rather than fight for Churchill. Since then it has never been so simple. In 1940 the French collaborators, choosing the path of nonviolence, made their peace with Hitler. A few years later the Jews of Europe went peacefully to their deaths at Auschwitz. Seeing what happened in France, I decided it was after all better to fight for England. Seeing what happened in Auschwitz, the surviving Jews decided it was better to fight for Israel. Nonviolence is often the path of wisdom, but not always. Love and passive resistance are wonderfully effective weapons against some kinds of tyranny, but not against all. There is a tribal imperative of self-preservation that compels us to use bullets and bombs against the enemies of the tribe when the tribe's existence is threatened. When it is a question of survival, passive resistance may be too slow and too uncertain a weapon.

Granted that the tribal imperative allows the members of the tribe to bear arms in self-defense, does this make Mutual Assured Destruction acceptable? Mutual Assured Destruction is the strategy that has led the United States and the Soviet Union to build enormous offensive forces of nuclear bombers and missiles, sufficient to destroy many times over the cities and industries of both countries, while deliberately denying ourselves any possibility of a defense. This strategy grew historically out of the strategic bombing doctrines of the 1930s, which were proved wrong in the war of 1939–45 against Germany but unfortunately gained an illusory success in the war against Japan. The basic idea of Mutual Assured Destruction is that the certainty of catastrophic retaliation will stop anybody from starting a nuclear war. It will indeed stop anybody who is cool and rational and in firm command of his own forces. If somebody is not cool and rational and not in firm command, what then? Then we trust to luck and hope for the best. If our luck turns sour, our missiles take off and carry out the greatest massacre of innocent people in all of history. I never have accepted this, and never will accept it, as either ethical or necessary.

Somewhere between the gospel of nonviolence and the strategy of Mutual Assured Destruction there must be a middle ground on which reasonable people can stand, a ground which allows killing in self-defense but forbids the purposeless massacre of innocents. For forty years I have been searching for this middle ground. I do not claim that I have found it. But I think I know roughly where it lies. The ground on which I will take my stand is a sharp moral distinction between offense and defense, between offensive and defensive uses of all kinds of weapons. The distinction is often difficult to make and is always subject to argument. But it is nonetheless real and essential. At least its main implications are clear. Bombers are bad. Fighter airplanes and antiaircraft missiles are good. Tanks are bad. Antitank missiles are good. Submarines are bad. Antisubmarine technology is good. Nuclear weapons are bad. Radars and sonars are good. Intercontinental missiles are bad. Antiballistic missile systems are good. This list of moral preferences goes flatly against the strategic thinking which has dominated our policies for the last forty years. And just because it goes against our accepted dogmas, it offers us a realistic hope of escape from the trap in which we are now ensnared.

Every soldier will quarrel with these moral distinctions, quoting the military maxim that offense is the best form of defense. In many cases the soldier's objection may be valid. It is often true that the best antitank weapon is a tank and the best antisubmarine weapon a submarine. Each case must be examined and judged on its merits. But in the larger view, there is no basic incompatibility between the demands of ethics and the realities of military operations. The rule that offense is the best form of defense should be a rule of tactics and not of strategy. It is a good rule for a battalion commander fighting a local battle but not for a commander in chief planning a war. It was this rule, extrapolated from the domain of tactics into the domain of grand strategy, that led both Napoleon and Hitler to disaster. So the moral distinction between defensive and offensive weapons would not forbid the use of tanks and aircraft in local counteroffensive operations. It would forbid the building of grand strategic forces of tanks and aircraft designed primarily for offensive warfare. And above all, it would forbid purely strategic offensive weapons, such as intercontinental missiles and missile-carrying submarines, for which no genuinely defensive mission is conceivable.

There, briefly stated, is my moral stand. I believe it is not in conflict with the ethics of a professional soldier who is honestly concerned to apply his skills to the defense of his country. Unfortunately, it is in conflict with the firmly held views of the civilian scientists and strategists who have led us down the road to Mutual Assured Destruction. The scientists have convinced the political leaders and the public that the supremacy of offensive weapons is an unalterable scientific fact. They have made the supremacy of the offensive into a dogma which the scientifically ignorant layman has no right to challenge. They argue that because the supremacy of the offensive is unalterable, the strategy of Mutual Assured Destruction is the best among the dismal alternatives that are open to us. But their basic dogma is in fact a falsehood. It is not true that defense against modern weapons is impossible. Defense is difficult, and expensive, and tedious, and complicated, and undramatic, and unreliable. But it is not hopeless. If we were to make the political decision to switch from an offense-dominated to a defense-dominated strategy, to redirect our weapons procurement and research and development, together with our diplomacy, toward the ultimate nullification of offensive weapons, there is nothing in the laws of physics and chemistry that would prevent us from doing so. We have drifted into the trap of Mutual Assured Destruction only because we have lacked the will and the moral courage to escape from it.

Why have our scientific strategists become so fanatically devoted to the doctrine of the supremacy of the offensive? The intellectual arrogance of my profession must take a large share of the blame. Defensive weapons do not spring like the hydrogen bomb from the brains of brilliant professors of physics. Defensive weapons are developed laboriously by teams of engineers in industrial laboratories. Defensive weapons are not academically respectable. Nobody would describe an antiballistic missile system with the phrase that Robert Oppenheimer used to describe the hydrogen bomb. *Defense is not technically sweet.*

One of the most tragic aspects of our situation is that it would have been much easier for us to switch to a defensive strategy in 1962 than it is now. The defensive strategy that I am advocating is not far removed from the strategy that I found recorded in the Soviet literature at the Arms Control and Disarmament Agency in 1962. If we had switched at that time, it would have required no great upheaval on the Soviet side to make the switch bilateral. Khrushchev was pushing hard the development of Soviet antiballistic missile systems and had deployed very few offensive intercontinental missiles. We had a chance then to offer Khrushchev a bilateral limitation of offensive forces to small numbers, leaving free the deployment of defensive systems which would in time have become adequate to nullify the limited offensive forces on both sides. Khrushchev, being at that moment ahead in defensive and behind in offensive weapons, would probably have accepted such an offer. At least we could have tried. We missed an opportunity that would not come twice.

In the fall of 1962 I went to England to attend a Pugwash meeting. Pugwash meetings are international gatherings of scientists who come together to discuss political and strategic matters in a friendly and informal fashion. A number of Russians were there, some of them politically knowledgeable and having close connections with their government. One of the Russians strongly implied, without saying it explicitly, that he would personally report our conclusions to Khrushchev when he returned home. The Russians knew that I worked for ACDA and incorrectly supposed that I was a good channel through which to convey information to my government. In private conversation they spoke to me in anguished tones, begging me to make the American government understand the urgency of the situation. They said that big decisions were soon to be made in the Soviet Union which would make the control of the arms race far more difficult. They gave me to understand that if there was to be any meaningful disarmament agreement in our lifetimes, it must be now or never. I have no doubt that they then knew that the tremendous build-up of Soviet offensive forces, whose true dimensions we would learn only many years later, was just about to begin. Unfortunately, I had no opportunity to deliver their message personally to John Kennedy. I spoke of it to my friends in ACDA, who did not take it seriously. And when I returned to ACDA the following summer we were preoccupied with the test ban.

The test ban was indeed a disastrous distraction. Just in those short Kennedy-Khrushchev years when there might have been political opportunities and willingness on both sides to consider drastic steps toward nuclear disarmament, people in responsible positions had no time to think about disarmament because they were too busy with the test ban. In the end, I too climbed on the test-ban bandwagon and missed my chance of making a serious effort to turn the arms race around. It is small consolation to reflect that by the summer of 1963 it was already too late to change the course of history. Within fifteen months, Kennedy would be dead and Khrushchev in unwilling retirement.

In the real world, the world in which human beings and nations have to live, the most important question about weapons is how they are used. Use of weapons is more important than production; production is more important than testing. The testing of weapons has only minor effects on human affairs, apart from the accidental rain of radioactive fallout that it causes. If we were serious in trying to regulate or abolish weapons of mass destruction, our order of priorities should be: use, production, testing. In ACDA and in the diplomacy of the Kennedy era, the priorities were exactly reversed. All of our scarce political capital was spent on the test ban. Hardly any attention was given to the way in which nuclear weapons were to be deployed and used. And yet, in the real world, arms races are driven by war plans and deployments. The basic reason we never succeeded in controlling nuclear weapons is that we never came to grips with the problem of use.

In 1959 George Kennan wrote an article with the title "Reflections on Our Present International Situation," which contained more wisdom than any other piece of writing that I have seen from that period. Kennan understood clearly what had to be done. He understood that we needed first of all to change our conceptions concerning the use of weapons, before we could hope to succeed in any technical approach to the problem of controlling the arms race. He had spent the greater part of his life in official dealings with the Soviet Union and understood the complexities of Soviet society. Kennedy appointed him ambassador to Yugoslavia and did not listen to what he had to say about larger issues.

Here is the gist of Kennan's message:

> Believe me, this commitment to the weapons of indiscriminate mass destruction which has dominated our strategic thinking, and increasingly our political thinking, in recent years, represents a morbid fixation of the most fateful and hopeless sort. No positive solution to any genuine human problem is ever going to be found this way. . . .
>
> My question, therefore, is: have we not had enough of this? . . . Let us remember that the Russians have been on record from the very beginning as favoring the total abolition of weapons of this nature. . . . I am assuming that we would not abandon our nuclear arms unless they did the same and unless adequate inspection facilities were granted. But would we be willing to do it even then? I have already mentioned the neglect of our conventional forces that has accompanied our increasing preoccupation with nuclear weapons. The concomitant of this weakness in conventional forces has been, as I understand it, a commitment to what is called the principle of first use of nuclear weapons: to their use, by us, in any serious military conflict, whether or not they are first used against us. This rests, of course, on the belief that we would be unable to look after our defense properly in contests where nuclear weapons were not used at all.
>
> I would submit that the first thing we have to do in order to put ourselves in a position to negotiate hopefully for an abolition of nuclear weapons, or indeed to have any coherent strategy of national defense, is to wean ourselves from this fateful and pernicious principle of first use. This means, obviously, a major strengthening of our conventional forces and, let us hope, of those of our allies. This is, I know, a disagreeable proposition. . . . It is, however, something that is wholly within our resources; what is lacking is only the will.

On February 4, 1961, there was a meeting of the council of the Federation of American Scientists in New York City. On the same day there was a major blizzard. Princeton was without electricity, and I had breakfast with my wife by candlelight before battling my way through the snow to the meeting. New York under a foot of snow had become suddenly friendly and beautiful. The meeting of the federation council was devoted to a long and careful discussion of the principle of first use of nuclear weapons. In the end we passed unanimously the following resolution:

We urge the government to decide and publicly declare as its permanent policy that the United States shall not use nuclear weapons of any kind under any circumstances except in response to the use of nuclear weapons by others. We urge that the strategic plans and the military deployments of the United States and its allies be brought as rapidly as possible into a condition consistent with the over-all policy of not using nuclear weapons first.

This statement was, so far as I know, the only public response of any kind to Kennan's appeal. On the following day the newspapers were filled with stories of the blizzard. Nobody printed any stories about No First Use. The federation never succeeded in making No First Use into a newsworthy political issue. The public does not want to think about No First Use. The public does not want to think at all about actual use of nuclear weapons.

During my two summers at ACDA I tried on various occasions to persuade my superiors that they ought to be devoting at least a little of their attention to the effects of our First Use policy on the possibilities of arms control. I was told emphatically that this was none of our business. The First Use policy was by that time deeply embedded in the structure of the NATO alliance, and therefore came under the jurisdiction of the State Department rather than ACDA. We in ACDA could not afford to antagonize the State Department by questioning the wisdom of its policies. If I wanted to raise awkward questions about the First Use policy, I had better dissociate myself from ACDA and do it somewhere else.

After Kennedy's death came the Vietnam years. First Use then became an even more frightening and immediate possibility. During those years I sometimes heard the subject discussed at meetings of government officials. At one such meeting, Official X expounded the United States First Use doctrine with an unimaginativeness worthy of Doctor Strangelove. He handed around copies of a memorandum entitled "Situations in Which the Use of Tactical Nuclear Weapons Is Plausible." The memorandum has no secret stamp on it. Number one on his list of situations was "Containing a Chinese Invasion with Minimum Risk of Accidentally Involving Russia." Official Y, sitting in the audience, scribbled a note on a piece of paper and passed it to me surreptitiously: "In other words, Nuke the Gooks and Polite the Whites." Official Y was one of the people in the Defense Department who were trying, all through the sad Vietnam years, to inject a voice of sanity into our military decisions. Those people were powerless either to stop the war or to change the style in which it was fought. All they could do, all they did do, was to keep the war from becoming an even greater disaster than it already was.

In 1966, at another such meeting, official Z said, "I think it might be a good idea to throw in a nuke now and then, just to keep the other side guessing." Hearing this, I was too astonished to protest. Official Z was, as it happened, not only impervious to argument but also deaf. It was impossible to be sure whether he was speaking in jest or in earnest. This was at a time when President Johnson was deliberately escalating the war without revealing his true intentions. All possibilities, including the possibility that Johnson would listen to Z's advice, had to be taken seriously. After the meeting ended, I checked with three other civilian scientists who were present, to make sure that Z had really said what I heard him say. They were all as shocked as I.

The four of us decided that something must be done. A formal protest against Z's remark would be completely ineffective. We concluded that the only way we might exert some real influence was to carry out a detailed professional study of the likely consequences if Z's suggestions were followed. We obtained permission from the Defense Department to make such a study. For three weeks we worked hard, collecting facts about the deployment of forces on both sides in Vietnam and analyzing the results of introducing nuclear weapons into the conflict. We carried through the analysis in a deliberately hard-boiled military style, and we summarized our conclusions in a report entitled "Tactical Nuclear Weapons in South-East Asia." Our analysis demonstrated that even from the narrowest military point of view, disregarding all political and ethical considerations, the use of nuclear weapons would be a disastrous mistake. We handed the report to our sponsors in the Defense Department. That was the last we saw of it.

I have no way of knowing whether anybody ever read our report. I have no way of knowing whether there was ever any real danger that Johnson would use nuclear weapons in Vietnam. All I know is that if Johnson had ever considered this possibility seriously and had asked his military staff for advice about it, our report might have been helpful in strengthening the voice of those who argued against it. All we could do was to buttress with some hard military facts the arguments that Johnson's advisers might have used to dissuade him.

This we did. I have no reason to believe that our report had the slightest actual effect on the course of the war in Vietnam. But it could conceivably have had an effect on the fate of mankind, far greater than the effect of anything I did at the Arms Control and Disarmament Agency.

The Defense Department, even in the worst days of the Vietnam war, was never monolithic and never totally intolerant of criticism. The majority of officials in the Pentagon were like X, conscientious and unimaginative. A few were like Y, actively critical and trying to push the department toward saner policies. A few were like Z, proving that the radical students' image of a Pentagon warmonger was not wholly unreal. A great deal depended on whether Y or Z prevailed over X in the decisions that were made from day to day. By coming in from the outside, encouraging Y and opposing Z, a scientist like me could hope to have some small but real influence on these decisions.

The most ambitious attempt by civilian scientists to intervene in the Vietnam war on a technical level was a project called the Barrier. The idea of the Barrier was to stop enemy soldiers from walking into South Vietnam, by means of an elaborate system of electronic burglar alarms and minefields dropped by airplanes along the frontiers. Secretary of Defense McNamara was enthusiastic about the Barrier, believing that it would be a substitute for the costly and politically unpopular use of American ground troops in search-and-destroy operations. The professional soldiers were less enthusiastic. They did not believe it would work. I was invited to join the Barrier project and considered with some care the ethical questions that it raised. According to my general principle of preferring defensive strategies, the Barrier was theoretically a good idea. It is morally better to defend a fixed frontier against infiltrators than to ravage and batter a whole country. But in this case, if one believed that the war was wrong from the beginning, a shift to a defensive strategy would not make it right. I refused to have anything to do with the Barrier, on the grounds that the ends it hoped to achieve were illusory. But I do not condemn my friends who worked on it in good conscience, believing that it would save many lives and mitigate the effects of the war on the civilian population of Vietnam. Their efforts were in vain, for the Barrier was never installed. If it had been installed, it would not have changed the course of history.

I believe that the Barrier would have been not only effective but morally good, if there had existed inside it a government and a people with the will and the competence to operate the system themselves. As a part of an indigenous effort of a country to defend itself, the Barrier would have made sense. What made no sense was for American technicians and air crews to operate a sophisticated defense system around a territory that had no political cohesion and no capable military forces of its own. It is unfortunate that the concept of the Barrier grew out of a hopeless attempt to save the American intervention in Vietnam from inexorable defeat. The association with Vietnam gave a bad name to a good idea.

In the long run, the survival of human society on this planet requires that one of two things happens. Either we establish some kind of world government with a monopoly of military power. Or we achieve a stable division of the world into independent sovereign states, with the armed force of each state strictly confined to the mission of defending its own territory. On humanistic and cultural as well as political grounds, I vastly prefer the second alternative. Fortunately, the majority of people seem to share my preference. From the beginnings of human history until today, great empires have tended to disintegrate and world-government movements have failed to attract wide public support. If we consider world government either undesirable or unattainable, then the aim of our military and diplomatic efforts should be, not to abolish nationalism, but to guide the forces of nationalism into truly defensive channels. We should strive to build a peaceful and harmonious society of independent nations, in which each country maintains a citizen army as Switzerland does now, posing no threat to its neighbors but ready to fight like hell against anybody who comes with dreams of conquest.

It is important for long-range stability that peaceful countries be well armed and well organized in self-defense. There will always from time to time be crazy demagogues like Hitler and technological surprises like the invention of gunpowder or nuclear weapons. Two factors, one technical and one human, make the long-range outlook hopeful for self-defense. The technical factor is the increasing effectiveness of small and sophisticated defensive weapons, precisely guided tank-killers and aircraft-killers and missile-killers, well suited to the defense of a fixed frontier. The war of 1973 in the Middle East gave only a foretaste of what these weapons can do. In the future we can, if we have the will to do so, negotiate arms control agreements that push the balance of technology still further to the advantage of defense. The human factor favoring self-defense is the invigorating effect of genuine political independence. Switzerland and Finland and Israel, countries which depend for defense upon their own efforts rather than upon alliances, have outstandingly efficient armies. Of all the countries that I have visited, these are the only ones in which a young man of good family who enjoys soldiering is not regarded as mentally subnormal.

We have a long way to go, from our present world of Mutual Assured Destruction with overwhelmingly large offensive forces to my dream world of independent countries efficiently defended by Swiss-style armies. How can we hope to get from here to there? I do not know. All I know is that we must get there, by one way or another, if we are to survive on this planet. If only we can all agree that our present situation is humanly and ethically unacceptable, we may find that the way to a better world is not as impassable as it seems.

The best hint I can find of a hopeful road into the future is a tale from the distant past. The perspective of a hundred and sixty years may help us to see clearly what are the features of an arms control agreement that give it durability. So I will briefly tell the story of the Rush-Bagot agreement, limiting naval armaments on the Great Lakes of North America. The agreement, made official in 1817 by Acting Secretary of State Richard Rush and the British Minister in Washington Sir Charles Bagot, stipulated as follows:

"The naval force to be maintained upon the American Lakes by His Majesty and the Government of the United States shall henceforth be confined to the following vessels on each side, that is: On Lake Ontario, to one vessel, not exceeding one hundred tons burden, and armed with one eighteen-pound cannon. On the Upper Lakes, to two vessels, not exceeding like burden each, and armed with like force. On the waters of Lake Champlain, to one vessel not exceeding like burden . . ." And so on.

The fleets deployed on the lakes in 1817 were much larger than the agreed limit, and the individual ships were too big to be sailed down the St. Lawrence River. The agreement required a substantial act of disarmament, which was promptly carried out by dismantling ships on both sides. The main objective of the agreement was to avoid confrontations which might lead to a renewed outbreak of the indecisive War of 1812. This aim was achieved. The agreement paid no attention at all to the problems of technological innovation. There is no sign in it that Mr. Rush and Sir Charles were disturbed by the thought that eighteen-pound cannon would not forever remain the last word in naval armament.

For a hundred years after the signing of the agreement, technological innovations were constantly creating difficulties in the implementation of it. During these years the American-Canadian frontier was not always as peaceful as it later became. In 1841 Britain violated the agreement with two steam frigates. In 1843 the United States responded with a ship of 685 tons carrying two six-inch guns whose shot could hardly have weighed less than eighteen pounds. And so it went on. From the 1840s onward there was never a time when one side or the other was not technically violating the agreement. As the political relations between the two sides gradually became less acrimonious toward the end of the nineteenth century, the magnitude of the violations increased. Each new violation was greeted with vehement protests from the other side, but as time went on the protests became less public and more ritualistic. In the year 1920 a senior official of the Canadian Navy was still writing, "It is . . . of the utmost importance that troops should be ready to immediately occupy the American shore of the St. Lawrence . . . and that a good supply of mines should be available in Canada for blocking the Straits of Mackinac, Detroit river, etc." But by that time nobody outside the military staffs was prepared to take such nightmares seriously.

The fact that the Rush-Bagot agreement was technically violated did not destroy its political usefulness. Through the worst periods of Canadian-American tension, the agreement was kept legally in force and was instrumental in holding these tensions in check. Political leaders on both sides found the agreement helpful, and used it effectively to pacify the bellicose elements on their own side of the border as well as to castigate the bellicose elements on the other side. The technical details of the agreement were important in 1817 but grew less and less important as its age and venerability increased. Now, after a hundred and sixty years, the agreement is still legally in force and is still technically violated several times every year. It has passed into folklore as a symbol of enduring peace.

I have a dream that a hundred and sixty years from now, some professor of physics will be looking back on the history of the treaty between the United States and the Soviet Union prohibiting deployment of bombers and missiles with nuclear warheads. He will, if all goes well, be explaining how the technological defects of the treaty did not prove to be fatal. He will explain how the treaty was technically violated by each of the great powers in turn during the turbulent first half of the twenty-first century. And how, in spite of flagrant violations, the treaty remained in force. And how, after the first demonstration of a cheap and effective non-nuclear antiballistic missile system by the Japanese, strategic offensive weapons gradually became obsolete and were retained only in small numbers for ceremonial purposes. That is, if we are as wise as Rush and Bagot. And if all goes well.

14

The Murder of Dover Sharp

I was wakened at six twenty-three in the morning of April 11, 1969, by a tremendous crash, followed by shouts of "Help!" I thought somebody must have driven a car smack into the Faculty Club at seventy miles per hour. I discovered then that I am, after all, a coward. Instead of running out immediately to the rescue, I dithered for about a minute, trying to pull myself together to face whatever had to be faced. For a minute I was paralyzed. And in that minute Dover Sharp burned to death.

During the Second World War and for many years afterward I used to have a recurrent nightmare. In my dreams I would see an airplane falling out of the sky. The airplane would crash and burst into flames near where I stood. I would stand there in terror, unable to lift my feet from the ground, and watch the people inside the airplane burning. I would strain and strain, trying to force myself to move, until I woke sweating and breathless in my bed. After that morning in Santa Barbara when Dover Sharp was murdered, the nightmares never came back.

Finally, I ran out of my room and down the stairs to the Faculty Club patio. I found that there had been no car crash. Two students were carrying Dover Sharp into an ornamental pool, which extinguished his burning clothes. He sat there in the pool and he did not look too bad. His legs were black and one hand was bleeding. I telephoned the hospital rescue squad, but they told me the explosion had already been reported and an ambulance was on its way.

After a few minutes the ambulance came. The students lifted Dover Sharp onto a stretcher and the ambulance crew took him away. A fire truck came next, and men with hand extinguishers quickly put out the fire that was burning in the dining room. We thought then that Dover Sharp would be all right. He had talked cheerfully with the students as they put him into the ambulance. But at noon we heard that he was burned so extensively that he was not likely to survive. He died in the hospital two days later.

The Faculty Club has only six bedrooms. Dover Sharp was caretaker of the building and lived in one of the bedrooms. I had come to the University of California at Santa Barbara as a visiting lecturer and was occupying another bedroom. Nobody else was in the building at the time of the explosion. Dover Sharp had come down in the morning and found a large cardboard box lying in front of the door that opened into the dining room. It was booby-trapped to explode when he opened it. It contained a half-gallon wine jug filled with gasoline, a six-inch piece of pipe packed with high explosive, and a battery-powered fuse to set it off. There was no message to indicate who had put it there or why.

The police investigating the murder called me in for questioning. I was not able to tell them anything useful. They did not ask me why I had been dithering in my room during the minute that it took the students to run across to the rescue from the San Rafael dormitory. For the police, that minute of delay had no bearing on the case. Only for Dover Sharp, it might have made the difference between life and death. And for me, it is a fact which I cannot change. I have to live with it as best I can.

The psychologist Robert Lifton has written a book, *Death in Life,* about the survivors of the atomic bombing in Hiroshima. He describes their feelings as told to him in interviews seventeen years after the bombing. Through all their stories runs the common thread of guilt for having lived when others died. Lifton quotes the words of Albert Camus, a survivor of the French resistance movement in the Second World War:

> In the period of revolution, it is the best who die. The law of sacrifice brings it about that finally it is always the cowardly and prudent who have the chance to speak since the others have lost it by giving the best of themselves. Speaking always implies a treason.

Frank Thompson left behind him a book of poems and letters. But all I have left of Dover Sharp is a name. Dover Sharp. At least I will hang on to that. I forget what he looked like, how his voice sounded, what words of greeting he used when I came down to breakfast at the Faculty Club. I hardly spoke to him all the time I was there. I never knew him as a person. I treated him as if he were part of the furniture. It is a bad habit that many professors have, to treat caretakers of buildings as if they were part of the furniture. This bad habit was one of the causes of Dover Sharp's death. If I had got to know him as a friend and as a human being, I would not have hesitated to save his life.

That spring was a time of turmoil in universities all over the United States. In Santa Barbara some of the student radicals had organized a "Free University" in the Student Center near the Faculty Club. I heard rumors that the Free University was offering courses in guerrilla warfare and in the manufacture of homemade weapons. Some of the professors were saying that the Faculty Club, with its elegant adobe-style architecture and its privileged clientele, had been chosen by the radicals as a suitable target for their resentments. But when I went into the Free University and gave talks there about the ethics of defense, the students there seemed as peaceful and friendly as those outside in the official university. There was a big poster on the wall with a newspaper account of Dover Sharp's death and the single word "WHY?" printed over it in huge letters. No evidence was ever found linking the radical students to the murder.

On the Sunday after Dover Sharp died, I looked out of the window of my room at the Faculty Club. It was a warm, sunny day. The blood and ashes had been washed away from the side of the ornamental pool where Dover Sharp had sat. Crowds of children were running and splashing in the pool just as if nothing had happened. Their parents were sunning themselves in the patio and discussing the state of the world. "Drive your cart and your plow over the bones of the dead," said William Blake. It is a hard saying, but there is much wisdom in it. I listened to the happy shouts of the children and wished that my own children were there too. I was thinking how lucky we all were, all of us except Dover Sharp, that this time it was only a gasoline bomb and not plutonium. Next time we would perhaps not be so lucky. I went down and sat in the patio so that I could be closer to the children.

Ted Taylor, ever since I met him for the first time in the little red schoolhouse in San Diego, had been obsessed with visions of nuclear weapons in the hands of terrorists. Whenever he had a chance to talk to me privately, when we were working together on Orion and when we saw each other occasionally in later years, he used me as a sounding board for his worries. For ten years, while nobody else in the world seemed to be worrying about the problems of nuclear terrorism, he worried. From his experience in Los Alamos, he knew better than anybody how easy it is, given a few pounds of plutonium, to build a bomb that can kill thousands of people or make a city uninhabitable. He worried about criminals stealing weapons ready-made, and he worried about criminals stealing plutonium and making their own weapons. He worried about international terrorist organizations long before the Baader-Meinhof gang and the Red Brigades became active. I was one of the few people to whom he could talk freely. Hour after hour we would sit together and examine every detail of the problem, discussing how and where plutonium might be stolen, how and where a small group of people might process the plutonium chemically and make it into a bomb, how powerful and how reliable such a bomb might be, how terrorists might use it to make threats of nuclear blackmail, and how a law-abiding society might organize its nuclear activities so that all these horrors might be avoided. I checked over the numbers, and Ted's arguments convinced me that it is indeed possible to imagine one or two people building a bomb in a private garage with only a minute fraction of the resources that were needed to do the job for the first time in Los Alamos. As I sat in the patio at Santa Barbara and watched the golden-brown children playing in the pool, I thought of Ted and his worries. If that cardboard box had had plutonium in it, the blood and ashes might not have been washed away so quickly.

Ted's awareness of the possibilities of nuclear terrorism presented him with an agonizing dilemma. On the one hand, he wanted to warn the authorities and the law-abiding public of the seriousness of the risk, so that simple precautions might be taken to make plutonium less easily accessible to criminals. On the other hand, if he were to call public attention to the problem, there was always a chance that he would be putting into the minds of terrorists possibilities that they would not have thought of by themselves. He knew that if he issued a public warning dramatic enough to command attention, and if an act of nuclear terrorism subsequently occurred anywhere in the world, he would feel himself to blame for it. Either way, whether he kept silent or spoke out, he was burdened with a terrible responsibility. At least a hundred times Ted and I thrashed over the arguments for silence and the arguments for speaking out. We never found any escape from the dilemma. For many years Ted remained silent. Then he decided to begin using private channels of communication to persuade the responsible officials in our government and in foreign governments to take better care of their plutonium. After that, if he failed in his efforts to alert the governments privately, he would again consider whether the time had come to take his message to the public.

In 1963, when the test-ban treaty was signed, Ted handed over the technical directorship of Project Orion to his second in command, Jim Nance, who gallantly steered the sinking ship through the final two years of its existence. Ted began a new career as Deputy Director of the Defense Atomic Support Agency, the branch of the Defense Department that had direct responsibility for taking care of nuclear weapon stockpiles. In that position he had excellent opportunities to find out how the United States government was handling its plutonium, and to discover weak points in the system where thieves might most easily break in. He also had opportunities to talk privately with senior officials of the Atomic Energy Commission and with important people in Congress. He spoke with these people and impressed on them how urgently necessary it was for them to begin mending the holes in their fences. He told them of case histories which he had himself quietly uncovered, of plutonium being stored and shipped in an appallingly casual fashion. His efforts were largely in vain. Two factors worked against him. First, the responsible officials were told by their own experts that nobody could build homemade bombs as easily as Ted imagined. Second, there were complicated jurisdictional snags that interfered with the setting up of uniform standards for protecting plutonium. Military plutonium, civilian government plutonium and industrial plutonium were handled by three different bureaucracies, and nobody had the authority to impose standards on all of them. After a while, Ted concluded that it was impossible, working quietly from the inside, to persuade the government authorities to take effective action. It was impossible even to convince them that they had a serious problem on their hands.

Although his attempt to alert the United States government had failed, Ted was still not ready to take the risk of alerting the public openly. He decided he must make one more effort at private persuasion, this time on an international level. The International Atomic Energy Agency, an organ of the United Nations with headquarters in Vienna, has the responsibility for establishing international standards and rules for the safeguarding of civilian nuclear activities. The IAEA standards are weak and not universally accepted, but at least they represent an international effort to hinder proliferation of nuclear weapons technology. So Ted moved to Austria. He resigned from his United States government job and settled with his wife and five children in Vienna. He had no official connection with IAEA and no financial security. He planned simply to stay in Vienna until the money ran out and see what he could do.

He stayed in Austria for two years and established lasting friendships with many of the technical people on the IAEA staff, with Indians, Russians, Yugoslavs and Western Europeans. He was able to convince many of them of the importance of tighter safeguards against nuclear theft. The technical people knew very well how many loopholes the IAEA standards left open. But Ted was less successful when he tried to talk to the political people at the upper levels of the IAEA administration. The political people told him that IAEA could do nothing without the approval of the member governments, and that the governments were in no mood to give IAEA any stronger policing powers than it already possessed. Ted returned to America feeling that his mission to Vienna had failed. The IAEA seemed as unwilling as the American government to contemplate any drastic moves that might be politically unpopular. But in fact his years in Austria had not been wasted. The wide and warm international contacts that he made there were to prove enormously helpful to him in later years.

The murder of Dover Sharp happened a few months after Ted returned from Vienna. Ted came out to Santa Barbara and spent a day with me. We were both despondent. I was mourning for Dover Sharp, and Ted had spent four of his best years fighting for nuclear safeguards in Washington and in Vienna without any visible result. We looked out at the world around us and saw incidents of random violence and terror becoming everywhere more frequent. We gloomily decided that the world was too stupid to learn anything from these small disasters. It seemed it was only a question of time before there would be a big disaster, an act of meaningless violence like the one in Santa Barbara but on a nuclear scale.

But Ted is stubborn. He continued his quiet campaign for nuclear safeguards in the United States, working this time through the Ford Foundation. The foundation gave him financial support for a thorough study of the problems of nuclear theft, to be done in collaboration with Mason Willrich. Willrich was on the staff of the Arms Control and Disarmament Agency when I was there. He is a lawyer by trade. Legal technicalities are as important as nuclear physics in the proper handling of safeguards. Willrich and Taylor made a good combination. Together they wrote a book, *Nuclear Theft: Risks and Safeguards,* which was published by the Ford Foundation in 1974.

When he decided to write the Ford Foundation book, Ted had finally made up his mind to tell the public what he knew. *Nuclear Theft* would be a detailed public statement of the dangers of nuclear terrorism. He hoped that the deliberately undramatic and low-keyed style of the book would lessen the risk that criminals would take their cue from it. He wanted his public warning to be as unsensational as possible. However, a chance encounter caused events to take a different course. Mason Willrich played tennis one day with the writer John McPhee. John McPhee was looking for a subject for his next *New Yorker* magazine article in the series "A Reporter at Large." Willrich said, "How about nuclear terrorism?" And so McPhee embarked on the article which grew into a full-length profile of Ted Taylor and was later published as a book with the title *The Curve of Binding Energy.* On the dust jacket the publishers added a subtitle, "A Journey into the Awesome and Alarming World of Theodore B. Taylor."

McPhee knew from the beginning that his book would be a shocker. He intended to scare the public, and he did. He wrote the book with his usual meticulous accuracy and attention to detail, including the detail of Ted's ideas about how easy it is for terrorists to build bombs. He talked with Ted for days on end. He also talked at length with Willrich and with me. McPhee and Ted were faced once again with the same dilemma that Ted and I had discussed so many times in the old days in San Diego. Did we dare take the responsibility for making these facts public? We thrashed a few times more through the same old arguments. In the end McPhee said, "Look. No matter what we do, this stuff is not going to stay secret much longer.

It is much better to get an accurate firsthand statement from you out in the open first, rather than wait for some secondhand mixed-up version to come out and confuse the issue." So Ted agreed to talk freely to McPhee, and McPhee accepted the responsibility for presenting Ted's words in such a fashion as to produce the maximum public impact.

McPhee's book appeared a year before the Ford Foundation book. McPhee's book made Ted immediately famous. To an important extent, it prepared the ground and created an audience for the Ford Foundation book. Without McPhee, Willrich and Taylor might have failed to attract any significant attention. McPhee's timing was exactly right. The public responded to his message, and the terrorists didn't. At least, not yet.

Nuclear Theft is a scholarly book, covering in detail and in depth the whole range of issues, legal and technical, that McPhee's dramatic statement had opened to public discussion. It is remarkable how influential a book can be, if it is written clearly and objectively and makes no attempt either to conceal or to exaggerate dangers. Not only in the United States but all over the world, Willrich and Taylor changed the way governments think about nuclear proliferation. After ten years of being disregarded as a crank, Ted found himself showered with invitations to testify before congressional committees and to advise foreign governments. Everywhere the political authorities recognized Ted as the man who could best tell them what to do, in a realistic and practical way, to tighten their safeguards. Slowly, belatedly, things have been done. Nuclear theft is not as easy now as it used to be. It is still, inevitably, easier than it ought to be.

Willrich and Taylor were successful in providing a foundation of factual information for political discussions of nuclear safeguards. Their main conclusions have not been seriously challenged either by pro-nuclear or by anti-nuclear advocates. Because of the existence of their book, it has been possible to maintain a rational discourse between the two sides in the discussion of safeguards against theft. Both sides agree more or less on the facts and can argue rationally about remedies. Unfortunately, in the other two areas of nuclear controversy, reactor accidents and nuclear waste disposal, no comparably objective books have been written. In the arguments about accidents and waste disposal, there is no Willrich-Taylor statement of facts agreed to by both sides; polemical statements abound and rational discourse is hard to find.

In 1976 Harold Feiveson, another alumnus of the Arms Control and Disarmament Agency, was teaching the course Public Affairs 452 at Princeton University. The subject of the course was "Nuclear Weapons, Strategy and Arms Control." The class consisted of twelve undergraduates: ten political science majors and two physics majors. One of the two physics majors was named John Phillips. The course was informal. The students were required to read extensively in the literature of arms control and write papers on subjects of their own choice. In class they gave oral accounts of what they had written, and argued the issues with Feiveson and with each other. Feiveson invited me to join the class as an observer. I was glad to accept. It was exciting to watch the students' knowledge and understanding grow from week to week. I read the students' essays and joined in their arguments. In the last two weeks of the course, the students divided themselves into a United States team and a Soviet team and negotiated a disarmament treaty. I was amazed to see how well they caught the spirit of their roles. The Soviet team became as zealous as any Soviet diplomats in the defense of Soviet security.

John McPhee's book and the Willrich-Taylor book were both on the reading list for the course. When the time came to choose topics for the final set of essays, John Phillips said he would like to do a paper on nuclear terrorism. He thought that he, being a physics major, would be the best qualified to do a careful study of Ted Taylor's ideas. He wanted to see for himself whether it was really true, as Taylor claimed, that a determined terrorist group with some stolen plutonium could build an atomic bomb. He asked me to be his supervisor in this investigation. I agreed to supervise him, but told him I would give him no help on technical details. I approved his project because it fitted in well with the general objectives of the seminar. The purpose of his exercise was mainly to educate the other students in the class concerning the seriousness of the nuclear safeguards issue. We had discussed nuclear terrorism in class, but the other students had insufficient scientific background to judge for themselves whether the danger of terrorist bombs was real. John Phillips would help them decide. I gave him references to books that he could find in the Princeton library, and talked with him twice about his general plan of work. Otherwise he was on his own.

After six weeks John gave an oral report to the class. I was aston-ished when I heard what he had done. Instead of treating his prob-lem as an academic exercise, he had gone out on his own initiative into the real world. He went to Washington and got hold of declas-sified Los Alamos reports, which contain far more information than the textbooks I had recommended to him. He picked up the tele-phone and called the chief of the explosives division at the factory where real bomb components are made. And so on. Everywhere he turned, people were delighted to cooperate and to feed him informa-tion. The class listened to his story in shocked silence. When he was finished, their reaction was summed up by Pam Fields, one of the political science majors, who said quietly, "Well, John, I am afraid we will have to put you away."

John's written paper consisted of two parts. One part was a sum-mary of the information he had obtained and how he had obtained it. The other was a rough sketch of a bomb design and an explanation of how it would work. The second part was not particularly startling. He had mastered quickly and competently the principles of shock-wave dynamics. But his sketch of a bomb was far too sketchy for the question "Would it actually explode?" to have any meaning. To me the impressive and frightening part of his paper was the first part. The fact that a twenty-year-old kid could collect so much information so quickly and with so little effort gave me the shivers. I read through his paper, awarded him an "A" grade, and told him to burn it. To my relief, the term ended in June and the paper attracted no public attention.

In October, quite suddenly, a storm of publicity broke over us. John was not responsible for starting the publicity. It began because an undergraduate who was working as a part-time correspondent for the *Trenton Times* talked with one of the students who had taken Public Affairs 452. Within a few days, wildly exaggerated stories about John's bomb were appearing in newspapers and magazines all over the world. John's face was on the cover of Sunday supplements from Philadelphia to Johannesburg, and even the staid *New York Times* came out with the headline: "Nations Beat Path to Door of Princeton Senior for His Atom Bomb Design." John showed a fine sense of responsibility in his handling of the situation. At the begin-ning he refused offers to appear on television. Only later, when the affair had blown up and was completely beyond our control, he agreed to appear on television and tried to explain to the public the dangers of nuclear theft. Being a gifted actor, he enjoyed the television shows and the fame and the fan mail. He enjoyed being invited to Paris to debate the issue of nuclear proliferation with French governmental and industrial officials on nationwide French television. But his head was not turned. Being a world celebrity was for him just another part of his education.

I, too, with one half of my mind, enjoyed the publicity. I especially enjoyed watching John on television and seeing how well he was able to put across to the audience the lessons he had learned in Public Affairs 452. But the other half of my mind was filled with fear and disgust. The media, as soon as they got hold of John's story, exploited it with little regard for truth and with absolutely no regard for public safety. They emphasized John's youth and charm and his rapid rise from obscurity to fame and fortune. The message that they were conveying to the public seemed to be: "All you have to do is build a bomb in your backyard, and you, too, can be rich and famous." John was himself horrified by the irresponsible sensation-mongering that surrounded him. This was exactly the kind of publicity, giving a false glamour to acts of violence and terror, that Ted had been afraid of when he hesitated for so many years to make his warnings about nuclear terrorism public.

For several weeks in the fall of 1976 my telephone was ringing constantly, with journalists and television people pestering me for stories about John Phillips. I learned to hate these people and the morbid fascination with which they ran after stories involving bombs and terrorists. Later, after the storm had subsided, I began to see that this fixation of the media upon acts of violence is not the fault of the media people alone. In their running after bombs and bloody horrors, the media are only reflecting the morbid tastes of the public. A fascination with violence lies somewhere deep in the hearts of all of us. At heart, we are not much better than the crowds which used to come to the Roman Colosseum nineteen hundred years ago to watch the gladiators hack each other to pieces.

For better or for worse, Ted Taylor's warnings of the dangers of nuclear terrorism have now been broadcast to the world in a language that everybody can understand. The fact that no gang of terrorists or crazy fanatics has yet appeared with a nuclear weapon should not make us complacent. We grown-up people are only over-grown children who still like to play with dangerous toys. Dover Sharp's murderers are still at large among us.

15

The Island of Doctor Moreau

> Not to go on all-fours; that is the Law.
>   Are we not Men?
> Not to suck up Drink; that is the Law.
>   Are we not men?
> Not to eat Flesh or Fish; that is the Law.
>   Are we not men?
> Not to claw Bark or Trees; that is the Law.
>   Are we not Men?
> Not to chase other Men; that is the Law.
>   Are we not Men?

Mutability of species was the great discovery of nineteenth-century biology. Darwin established the fact that all species, including the human, change with time. Darwin knew well what distress his discovery would cause to people of conscience, and on that account delayed his publication of it for twenty years. He had no wish to emphasize the conflict between the idea of mutability of species and ordinary human values and feelings. The depths of that conflict were first explored by H. G. Wells in two works of macabre imagination, *The Time Machine* and *The Island of Doctor Moreau*. Wells was a writer of genius who also happened to be a trained biologist. He understood better than most of us the comedy of the individual human being, and yet he never lost sight of his biological background, of the human species emerging from dubious origins and groping its way to an even more dubious destiny. He published *Doctor Moreau* in 1896, soon after *The Time Machine* had made him famous. Both stories were profoundly antagonistic to the prevailing

atmosphere of late-Victorian optimism. Only later, when pessimism
became fashionable, did Wells become an optimist. He always liked
to swim against the tide. Long after the swings of fashion which first
acclaimed and then rejected Wells's optimistic writings, *Doctor Moreau* remains a classic. The island of beasts butchered into a sem-
blance of humanity by a mad physiologist is one of the most durable
nightmares in the literature of scientific horror.

Doctor Moreau has not only carved his beasts physically into
human form by plastic surgery; he has also forced their minds into
human patterns of behavior by incessant repetition of his Law. Gath-
ered in their squalid hut, they chant together, "Not to go on all-fours;
that is the Law. Are we not Men?" But that is not the worst of it. After
the chanting of the Law comes the hymn of praise to their Creator:

> His is the House of Pain.
> His is the Hand that makes.
> His is the Hand that wounds.
> His is the Hand that heals.
> His is the lightning-flash.
> His is the deep salt sea.
> His are the stars in the sky. . . .

With this hymn of praise, Wells raised the question that must ultimately be faced by all believers in scientific progress. Can man play
God and still stay sane? Wells did not ask or answer the question
explicitly. He was first of all a novelist, not a philosopher, and so he
let his story ask the question for him. The character of Doctor Moreau answers it with a resounding No.

Wells's hero, after escaping from the horrors of the island, comes
back to civilization like Swift's Gulliver, still haunted by what he has
seen, and alienated from his human kindred.

For several years now, a restless fear has dwelt in my mind, such a restless
fear as a half-tamed lion-cub may feel. My trouble took the strangest form.
I could not persuade myself that the men and women I met were not also
another, still passably human, Beast People, animals half-wrought into the
outward image of human souls; and that they would presently begin to
revert, to show first this bestial mark and then that. . . . And even it seemed
that I, too, was not a reasonable creature, but only an animal tormented with
some strange disorder in its brain, that sent it to wander alone, like a sheep
stricken with the gid.

Here we have, expressed with the personal touch that is Wells's hallmark as a writer, the anguish of every human being who faces in his imagination the implications of modern biology. The progress of biology in general, and the mutability of species in particular, threaten to deprive mankind of two psychological anchors: our sense of our own identity, and our sense of brotherhood one with another. The uniqueness of the human species, and the brotherhood of mankind: these are two anchors that may be essential to our sanity. Whoever has visited Doctor Moreau's island has lost these anchors. He will never again be sure what manner of creature he is.

We have come a long way since 1896. We have understood, to an extent that Wells in his wildest dreams never imagined, the language of the DNA molecules in which the instructions for reproducing living creatures are written. Our understanding is still fragmentary and partial. But it can hardly take us more than a few decades, or at most a century, to decipher and read the DNA language in all its details. We shall soon understand not only the alphabet and the words of that language, but the syntax and the paragraphs, the complete pattern of organization that enables a few molecules of DNA to tell an undifferentiated egg cell how to divide and grow into a human being. And at that point Wells's old nightmare comes again to haunt us. When we have learned in all detail how life is reproduced, we shall also have learned how life is produced. Whoever can read the DNA language can also learn to write it. Whoever learns to write the language will in time learn to design living creatures according to his whim. God's technology for creating species will then be in our hands. Instead of the crude nineteenth-century figure of Doctor Moreau with his scalpels and knives, we shall see his sophisticated twenty-first-century counterpart, the young zoologist sitting at the computer console and composing the genetic instructions for a new species of animal. Or for a new species of quasi-human being. Then Wells's question will have to be answered, not in a science fiction story but in our real world of people and governments. Can man play God and still stay sane? In our real world, as on the island, the answer must inevitably be no.

Wells was right in seeing the long-range threats to human sanity and human survival coming from biology rather than from physics. The hydrogen bomb can easily destroy our civilization but can hardly exterminate us as a species. The hydrogen bomb is almost a simple problem, compared with the problems posed by a deliberate distortion or mutilation of the genetic apparatus of human beings. Nuclear war is not the worst of imaginable horrors. Doctor Moreau's island is worse.

After Wells, the next biologist who gazed into the future to see the shape of things to come was J. B. S. Haldane. Haldane published in 1924 a little book, *Daedalus, or Science and the Future,* which is in many ways the best book ever written about the human consequences of progress in biology. Haldane has a lighter and more ironic style, but his conclusions are no less bleak than those of Wells. Most of the biological inventions which Aldous Huxley used a few years later as background for his novel *Brave New World* were cribbed from Haldane's *Daedalus.* Haldane's vision of a future society, with universal contraception, test-tube babies, and free use of psychotropic drugs, became a part of the popular culture of our century through Huxley's brilliant dramatization. Huxley added to Haldane's picture an important new twist, the manufacture of large batches of identical human beings by cloning. But in its essence, Huxley's Brave New World is only Moreau's Island enlarged and brought up to date by the addition of modern technology. Drugs replace whips and genetic programming replaces surgery. Huxley's hero, like Wells's, is a natural man totally disoriented by the discovery that the fellow creatures with whom he tries to form human relationships are not fully human. To a person with truly human sensibilities, Huxley's world of synthetic happiness is as alien as Wells's island of misery and degradation.

Haldane did more than add technical sophistication to Wells's nightmare. He also presented a new vision of the character of the scientist. Doctor Moreau was a pathological character of a simple type: a man of great intellect driven crazy by frustrated ambition. Haldane chooses for his archetype of the experimental biologist the mythical figure of Daedalus, who according to legend superintended the successful hybridization of woman and bull to produce the Minotaur.

The chemical or physical inventor is always a Prometheus. There is no great invention, from fire to flying, that has not been hailed as an insult to some god. But if every physical and chemical invention is a blasphemy, every biological invention is a perversion. . . . I fancy that the sentimental interest attaching to Prometheus has unduly distracted our attention from the far more interesting figure of Daedalus. He was the first to demonstrate that the scientific worker is not concerned with gods. The unconscious mind of the early Greeks, who focussed in this amazing figure the dim traditions of Minoan science, was presumably aware of this fact. The most monstrous and unnatural action in all human legend was unpunished in this world or the next. Socrates was proud to claim him as an ancestor. . . .

We are at present almost completely ignorant of biology, a fact which often escapes the notice of biologists, and renders them too presumptuous in their estimates of the present position of their science, too modest in their claims for its future. . . . The conservative has but little to fear from the man whose reason is the servant of his passions, but let him beware of him in whom reason has become the greatest and most terrible of the passions. These are the wreckers of outworn empires and civilizations, doubters, disintegrators, deicides. . . . I do not say that biologists as a general rule try to imagine in any detail the future applications of their science. They do not see themselves as sinister or revolutionary figures. They have no time to dream. But I suspect that more of them dream than would care to confess it. . . .

The scientific worker of the future will more and more resemble the lonely figure of Daedalus as he becomes conscious of his ghastly mission and proud of it.

> Black is his robe from top to toe,
> His flesh is white and warm below,
> All through his silent veins flow free
> Hunger and thirst and venery,
> But in his eyes a still small flame
> Like the first cell from which he came
> Burns round and luminous, as he rides
> Singing my song of deicides.

Haldane evidently fancied himself a Renaissance man, classical scholar and poet as well as biologist. His portrait of Daedalus is in its way as impressive as Goethe's portrait of Faust. But does all this poetic imagery have anything to do with reality? Do our professors of biology nowadays ride around their laboratories singing songs of deicides? Obviously not, in the literal sense. In their outward appearance, professors of biology resemble Daedalus just as little as professors of physics resemble Faust. And yet, on a deeper level, the legends speak truth. Teller, with his indomitable urge to light a thermonuclear fire on earth, was following in the footsteps of Faust. Darwin, quietly accumulating fact upon fact until he was ready to demolish forever the comfortable universe of Victorian piety, was a deicide as implacable as Daedalus. The modern molecular biologists who are learning to read and write the language of the genes will in the end, whether they intend it or not, demolish our comfortable world of well-defined species with its impassable barriers separating the human from the nonhuman. In each of them the spirit of Daedalus is riding.

Two things we have learned from Wells and Haldane. Man cannot play God and still stay sane. And the progress of biology is inescapably placing in man's hands the power to play God. But from these two facts it does not follow that there is no hope for us. We still can choose to be masters of our fate. To deny to any man the power to play God, it is not necessary to forbid him to experiment and explore. It is necessary only to make strict laws placing the applications of his knowledge under public control. Such laws already exist in many countries, restricting the use of dangerous medical procedures, drugs and explosives. In the future, we shall need to arrive at a reasonable political compromise, allowing biologists freedom to explore the marvels of genetic programming that underly the living world, while severely limiting the right of anyone to program new species and let them loose where they may disturb nature's balance or our own social equilibrium. Such a political compromise should not be impossible to maintain. The biologists have already made a good beginning.

The biologists showed extraordinary wisdom in their handling of the problem of biological weapons. Their wisdom has greatly improved our chances of finding acceptable political solutions to the problems of regulating other possible abuses of biology. Aldous Huxley in *Brave New World* mentions in passing the anthrax bombs with which human populations were exterminated in the Nine Years' War preceding the establishment of the benevolent dictatorship of the World Controllers. Anthrax bombs are a real possibility. They could be cheap and easy to manufacture and extremely lethal to unprepared populations. Anthrax bacilli are peculiarly unpleasant because they form spores which survive and remain infective for many years. Designers of biological weapons have generally preferred to use other types of disease germ, which are as lethal as anthrax but not as persistent. If any of these weapons were ever used on a large scale, they would probably cause as much death and human misery as a war fought with hydrogen bombs.

It stands to the everlasting credit of the international fraternity of biologists that biologists, with rare exceptions, never pushed the development of biological weapons. Also, biologists persuaded the governments of those countries that had started serious biological weapons programs to abandon their programs and to destroy their stockpiles of weapons. To measure the greatness of the biologists' achievement, we may imagine what the world would now be like if the physicists had first declined to push nuclear weaponry and later persuaded their governments to destroy nuclear stockpiles. The biologists, unlike the physicists, came through their first trial at the bar of history with clean hands.

The man who did more than any other single person to rid the world of biological weapons is Matthew Meselson, professor of biology at Harvard. He came as I did to the Arms Control and Disarmament Agency for the summer of 1963, to see what he could do for peace. Unlike me, he did not allow himself to be distracted by the excitements of the test-ban negotiations, but kept to his own business. His business was biological weapons.

Meselson knew little about biological weapons when he came to ACDA. Like other academic biologists, he had had almost no contact with the closed military world in which biological weapons were developed and their uses were planned. Through ACDA he was able to gain access to that world. He talked with army officers who specialized in biological warfare, and read their writings. He moved freely in the world of biological agents and distribution systems. What he saw there appalled him.

The most frightening of all the things which Meselson discovered during that summer at ACDA was Army Field Manual 3-10. This was a booklet issued to combat units to instruct them in the details of biological warfare. A series of graphs is presented which tell how many biological-agent bomblets an aircraft should drop to cover a given area under given conditions, daytime or nighttime, for various types of terrain and various types of human target. The text is written in the same matter-of-fact prose that the army would use for a field manual on the proper method of digging a latrine. And the booklet is unclassified. It was, in 1963, widely distributed among United

States units and easily available to foreign intelligence services. It carried a clear message to any foreign general staff officers who might happen to read it. It said that the United States was equipped and prepared for biological warfare, that this was the way a modern army should be trained, that every country which wanted to keep up with the Joneses must have its own biological agents and its bomblets too.

After he read Field Manual 3-10, Meselson vowed that he would fight against this nonsense and not rest until he had got rid of it. He worked indefatigably, in private and in public, to expose the idiocy of American policies concerning biological warfare. His arguments rested on three main points. First, biological weapons are uniquely dangerous in providing opportunities for a small and poor country, or even for a group of terrorists, to do grave and widespread damage to a large country such as the United States. Second, the chief factors increasing the risk that other countries might acquire and use biological weapons are our own development of agents and our own propaganda as typified by Field Manual 3-10. Third, biological weapons are uniquely unreliable and therefore inappropriate to any rational military mission for which the United States might intend to use them, even including the mission of retaliation in kind for a biological attack on our own people.

Meselson found that it was not difficult to persuade military and political leaders to agree with his first two points. The crucial question was the third one. Did there exist any realistic military requirement for United States biological weapons? Here there was a division of opinion between the biological warfare generals and the rest of the military establishment. The biological warfare generals sincerely believed that we needed biological weapons to deter by threat of retaliation the use of biological weapons by others. Meselson had to show that their belief was based on an illusion. He appeared to confront them when they came to argue for their programs before congressional committees. He asked them, in his quiet and polite voice, "General, we would like to know, supposing that the United States had been attacked with biological weapons and the President had given the order to retaliate, just what would you do? Where, and how, and against whom, would you use our weapons?" The generals were never able to give him a clear answer. There was in fact no answer to these questions. Biological weapons are so chancy, their effects so unpredictable and uncontrollable, that no responsible soldier would want to use them if he had any available alternative. For the mission of retaliation in reply to a massive and deliberate biological attack, the alternative of nuclear weapons was available and would be preferred. After listening to Meselson's questions and to the generals' answers, the congressmen became convinced that his third point was valid. Even from the narrowest military point of view, our biological weapons policy made no sense.

In 1968 fate placed a great opportunity in Meselson's hands. Henry Kissinger had been for many years a Harvard professor, working in the building next door to Meselson's laboratory, and had followed the progress of Meselson's campaign against biological weapons. In 1968 Kissinger became right-hand man to President Nixon. Meselson urged Kissinger to move fast. Biological weapons were the one area in which Nixon could halt an arms race by unilateral action, with the assurance that Congress would support him. Kissinger and the other members of the National Security Council presented Nixon with the arguments for and against biological weapons. In November 1969, less than a year after taking office, Nixon announced the unilateral abandonment by the United States of all development of biological weapons, the destruction of our weapon stockpiles, and the conversion of our biological warfare laboratories to open programs of medical research.

This was Nixon's finest hour. It was a historic and statesmanlike action, fortunately completed before the shadows of Watergate began to close around him. It was a bold step, to undertake a major act of disarmament unilaterally. Many people in the government were saying, "Let us by all means get rid of biological weapons, but let us not do it unilaterally. Let us negotiate with the Russians and keep what we have until they agree to destroy theirs too." Meselson insisted that unilateral action must come first, negotiation second. If Nixon had begun with negotiations, there would have been endless discussions about the technical problems of monitoring violations of an agreement, with the probable result that no agreement could have been reached. At the very best, it would have taken years to negotiate a treaty, and in the meantime the biological weapons programs might have gained political support which would have made a treaty difficult to ratify. Nixon's unilateral action removed all these difficulties. After announcing the American decision to abandon biological weapons, Nixon invited the Soviet Union to negotiate a convention to make the action multilateral. Negotiations were begun, with the United States negotiating "from a position of weakness," having nothing more to give in exchange for Soviet compliance. According to orthodox diplomatic doctrine, to negotiate from a position of weakness is a mistake. But in this case the tactic was successful. The Soviet political leaders were evidently convinced by Nixon's action that their own biological weapons were as useless and as dangerous as ours. Brezhnev signed the convention, agreeing to dismantle his programs, in the summer of 1972, just nine years after Meselson arrived at ACDA and began to read Field Manual 3-10. Seldom in human history has one man, armed only with the voice of reason, won so complete a victory.

Meselson does not regard his victory as complete so long as chemical weapons are not also outlawed and abandoned. He has continued the fight against chemical warfare. In 1970 he traveled to Vietnam to investigate and document the use of chemical agents there. His arguments against chemical warfare are based on detailed knowledge of military history and doctrine. His case against chemicals is as robust as his case against biologicals. But to win the second battle will take him a little longer.

It is possible to imagine Meselson's tactics being used successfully against other varieties of dangerous weapons, and in particular against tactical nuclear weapons. Perhaps we could hammer at the tactical nuclear generals as adroitly as Meselson hammered at the biological generals, asking them, "Please, General, will you be so kind as to tell us, supposing that the North Koreans were overrunning Seoul and the South Koreans were in retreat, what precisely would you do? Where, and how, and against whom, would you use our nuclear weapons?" Perhaps the generals would be unable to find convincing answers to such questions. Perhaps we might conclude, after hearing their answers, that a unilateral withdrawal of tactical nuclear weapons would be to everybody's advantage.

Saving the world from biological warfare was for Meselson only a hobby. Through all those years he also pursued a successful career in biological research. He runs a laboratory at Harvard in which he explores the structure of genes. In his genetic research he uses various techniques, including the technique of cloning DNA molecules by artificial recombination. This "recombinant DNA" technique places a piece of DNA, from any gene which we desire to study, into a convenient bacterium, so that the multiplication of the bacterium produces a cloning of the gene. As a result of his leadership in the Harvard work with recombinant DNA, Meselson found himself embroiled in yet another political battle. The mayor of Cambridge, supported by a few distinguished biologists and by the radical faction of the Cambridge academic community, tried to prohibit experiments with recombinant DNA in Cambridge. Meselson and his colleagues told the mayor that they were doing nothing that endangered the health of the public. The city council voted to appoint a Citizens' Committee, a group of eight people unconnected with biological research, to advise the city whether or not to allow work with recombinant DNA to continue. Experiments were banned while the Citizens' Committee was studying the problem. The temporary ban lasted for seven months.

The Cambridge Citizens' Committee worked hard and heard all sides of the argument surrounding recombinant DNA. Meselson and his colleagues presented to the committee members the case for continuing recombinant DNA experiments. With inexhaustible patience, Meselson explained the many difficult technical and moral issues that the committee had to consider. The committee members listened to him and trusted his quiet uncertainty more than they trusted the loud certainty of his opponents. In the end they voted unanimously to recommend to the City of Cambridge the continuation of recombinant DNA experiments, subject to reasonable restrictions and supervision by local public health authorities. The city council accepted the committee's recommendations and Meselson could go back to work in his laboratory.

Why is there this intense public furor over recombinant DNA experiments? The public concern arose because two quite separate issues became confused. On the one hand, there may be an immediate danger to public health if certain kinds of recombinant DNA are grown in the laboratory and released into the environment in an irresponsible fashion. On the other hand, there are the long-range horrors, beginning with Doctor Moreau and ending with the cloning of human beings, that may come to pass as a result of misapplication of biological knowledge. The biologists who began the recombinant DNA experiments were aware of the possibility of an immediate public health danger. The molecular biologist Maxine Singer, wife of the Daniel Singer who had been the Federation of American Scientists' general counsel, published a statement calling attention to the danger, soon after the first experiments were done. In 1975 an international meeting of biologists voluntarily drew up a set of guidelines, prohibiting experiments that seemed to them irresponsible and recommending containment procedures for permissible experiments. Guidelines similar to theirs have now been accepted by biologists and governments all over the world. These guidelines have made any immediate public health hazard resulting from DNA experiments very unlikely. One cannot say that the immediate hazards are nonexistent, but they are smaller than the hazards associated with the standard procedures for handling disease germs in clinical laboratories and in hospitals. So from the point of view of the public health authorities, the risks of recombinant DNA experiments are adequately controlled. Why, then, is the public still scared? The public is scared because the public sees farther into the future and is concerned with larger issues than immediate health hazards. The public knows that recombinant DNA experiments will ultimately give the biologists knowledge of the genetic design of all creatures including ourselves. The public is rightly afraid of the abuse of this knowledge. When the National Academy of Sciences organized a meeting in Washington to give all sides of the recombinant DNA debate a chance to be heard, the public appeared in the guise of a gang of young people carrying placards and chanting, "We won't be cloned." The public sees, behind the honest faces of Matthew Meselson and Maxine Singer, the sinister figures of Doctor Moreau and Daedalus.

Recombinant DNA experiments are continuing in many places with great success. No harmful effects on the health of humans, animals or plants have been detected. But this does not mean that the long-range dangers of biological knowledge have vanished. Recombinant DNA is only one technique among many in the broad advance of biology. With or without recombinant DNA, the advance of biology will continue. It is biology itself, and not any particular technique, that is leading us swiftly onward into that uncharted ocean where Doctor Moreau's island lies. Matthew Meselson's purpose as a biologist and as a citizen is "to build an ethos for the future, one that says a deep knowledge of life processes must be used only to reinforce what is essentially human in us."

16

Areopagitica

In the fall of 1976, while the Cambridge Citizens' Committee was still at work, Princeton University asked the Princeton municipal authorities for permission to build two laboratories equipped for work with recombinant DNA. The Princeton municipalities were unprepared to make a decision and so followed the example of Cambridge. They appointed a Citizens' Committee to advise them. Our committee consisted of eleven citizens, of which I was one. Like the Cambridge committee, we worked hard for four months. Unlike the Cambridge committee, we were not able to produce a unanimous report. In the end we split eight to three, the majority saying yes to Princeton University, the minority saying no. We wrote separate majority and minority recommendations. But in spite of our differences of opinion, or rather because of our differences of opinion, my service on the Citizens' Committee was one of the happiest and most rewarding experiences of my life. We were struggling with deep problems and we became firm friends.

Our committee was a good cross section of Princeton. We were six men and five women, nine white and two black, four talkative and seven quiet. We had two medical doctors, two scientists, two writers, two teachers, a Presbyterian minister, an undersea photographer, and a retired lady who is a leader of the black community. Wallace Alston, the minister, Susanna Waterman, the photographer, and Emma Epps, the black community leader, were the unshakable minority. From the beginning it was clear that these three were the strongest characters on our committee and had the deepest convictions. I spent most of my time and effort in getting to know these three, understanding the philosophical roots of their objections to recombinant DNA, and attempting to find a compromise between their opinions and mine. In the end we knew that there could be no compromise, but our respect and liking for one another grew stronger as the hope of agreement faded.

The charge from the municipalities to our committee said clearly that we were to address our recommendations to the immediate problem of public hazards arising from recombinant DNA experiments in Princeton. The two doctors on the committee wanted to interpret our charge narrowly. Accustomed in their daily lives to balancing risks of life and death, they were impatient of lengthy discussions of remote contingencies. Judging by the standards of normal medical practice, they concluded that the public health hazards of recombinant DNA were well controlled by existing guidelines, and that this was all that our committee needed to say. They did not wish to waste their time arguing about broader philosophical issues. I felt great sympathy for the two doctors, busy people with heavy responsibilities, listening hour after hour to meandering conversations that they considered irrelevant.

On the other side, the minority of three felt even more strongly that it was wrong for us to confine our thinking to the immediate public health issues. For them it was a matter of conscience. They could not in good conscience shut out from their decision-making the great questions of human destiny to which recombinant DNA research is leading. I felt great sympathy for them too. Susanna Waterman summarized their position in the last sentence of her minority statement:

Based on the extraordinary and profound future impact of recombinant DNA research and its application within our delicate and finite biosphere, any decision to go forward with such research, if it is to go forward, should be firmly based on an informed public consent, on firm scientific data, and on democratic procedures.

Emma Epps, who celebrated her seventy-sixth birthday at one of our meetings, added to the minority report a briefer and more eloquent statement of her own:

My conscience tells me to say No to this, and I don't want to go against my conscience. Also, friends who are scientists say they don't see any reason why I should go against my conscience.

I am proud to be numbered among her friends.

In the end, although I felt personally and philosophically closer to the minority, I voted with the majority. I did so on legal grounds. From a legal point of view, the municipality of Princeton has a right and a duty to restrict any research at Princeton University that may cause a hazard to the health of citizens. But no public authority should have a legal right to restrict research merely because the people in positions of authority are philosophically opposed to it. Even though I accept the wisdom of the various philosophical misgivings that caused Alston, Waterman and Epps to vote no, I cannot accept the notion that the Borough of Princeton should have the power to impose their philosophical views upon Princeton University by municipal ordinance. As Thomas More says in Robert Bolt's play *A Man for All Seasons,* "I know what's legal, not what's right. And I'll stick to what's legal."

In June 1977 we presented our majority and minority recommendations to an unhappy borough council. The councillors had wanted us to tell them what to do. Since we spoke with a divided voice, they found themselves obliged to examine the issues in detail and to accept the responsibility for making a decision. They faced a long winter of studying recombinant DNA in addition to their normal responsibilities for municipal sewage and zoning variances. It took them nine months to make up their minds. During the nine months, Princeton enjoyed the distinction of being the only place in the world where recombinant DNA research was forbidden. Finally, in the spring of 1978, they voted five to one to accept the recommendations of our majority. An ordinance was passed, as in Cambridge, subjecting biohazardous research to municipal supervision. Democracy, in its slow and stumbling fashion, resolved a difficult and emotional issue, and still allowed the minority to feel that its views had been carefully weighed and not arbitrarily overridden.

As a reward for serving on the Princeton Citizens' Committee I was invited to Washington to testify at hearings of the Subcommittee on Science, Research and Technology of the U.S. House of Representatives. The subcommittee, with Congressman Ray Thornton of Arkansas as chairman, was making a serious effort to educate itself concerning the broader issues of national policy raised by recombinant DNA. Other committees of the House and Senate were studying the immediate problems of regulating biohazardous experiments. Ray Thornton wanted to take a longer view, to examine what the recombinant DNA debate might portend for the future relationships between science and government. His invitation to testify gave me a chance to make the voice of John Milton heard in Washington, as it had been heard long ago in London, speaking truth to power.

It has sometimes been said that the risks of recombinant DNA technology are historically unparalleled because the consequences of letting a new living creature loose in the world may be irreversible. I think we can find many historical parallels where governments were trying to guard against dangers that were equally irreversible. I will describe briefly two such historical parallels and leave you to decide for yourselves whether they throw light on our present problems.

My first example is the personnel security system that was set up by the U.S. Atomic Energy Commission in the years after the Second World War to protect atomic secrets. The government rightly decided that the consequences of letting atomic secrets loose in the world were irreversible and highly dangerous. The personnel security system was designed to provide the highest degree of containment for important secrets. Unfortunately, the regulations were so strict and the administration of them was so inflexible that the whole system came to be regarded by many scientists with some degree of contempt. As you all know, in 1954 Robert Oppenheimer came into collision with the officials whose job was the zealous enforcement of the rules. There was a battle, and Oppenheimer lost. I am not arguing that Oppenheimer was right. He did indeed behave arrogantly and irresponsibly toward the security officials. I am arguing that the Atomic Energy Commissioners, by the way they treated Oppenheimer, lost the respect of a great part of the scientific community. I believe further that the lasting alienation that resulted between the Atomic Energy Commission and the scientific community has been a major contributory cause of the difficulties that the nuclear enterprise has encountered in the last decade. So I advise you to watch out when you write the rules governing research with recombinant DNA. Write the rules flexibly and enforce them humanely, so that when some biologist, as brilliant and as arrogant as Oppenheimer, tries to set himself above the rules, he may not be perceived by his colleagues and by the public as a hero.

My second example is taken from a far more remote past. Three hundred and thirty-three years ago, the poet John Milton wrote a speech with the title "Areopagitica," addressed to the Parliament of England. He was arguing for the liberty of unlicensed printing. I have collected a few passages from his speech which speak to our present concerns. I am suggesting that there is an analogy between the seventeenth-century fear of moral contagion by soul-corrupting books and the twentieth-century fear of physical contagion by pathogenic microbes. In both cases, the fear was neither groundless nor unreasonable. In 1644, when Milton was writing, England had just emerged from a long and bloody civil war, and the Thirty Years' War which devastated Germany had still four years to run. These seventeenth-century wars were religious wars in which differences of doctrine played a great part. In that century, books not only corrupted souls but also mangled bodies. The risks of letting books go free into the world were rightly regarded by the English Parliament as potentially lethal as well as irreversible. Milton argued that the risks must nevertheless be accepted. Here are four of the salient points of his argument. I ask you to consider whether his message may still have value for our own times, if the word "book" is replaced by the word "experiment."

First, Milton was willing to suppress books that were openly seditious or blasphemous, just as we are willing to ban experiments that are demonstrably dangerous.

> "I deny not but that it is of greatest concernment in the Church and Commonwealth, to have a vigilant eye how books demean themselves as well as men, and thereafter to confine, imprison, and do sharpest justice on them as malefactors. I know they are as lively, and as vigorously productive, as those fabulous dragon's teeth, and being sown up and down, may chance to spring up armed men."

The important word in Milton's statement is "thereafter." Books should not be convicted and imprisoned until after they have done some damage. What Milton objected to was prior censorship, that books would be prohibited even from seeing the light of day.

Next, Milton comes to the heart of the matter, the difficulty of regulating "things uncertainly and yet equally working to good and to evil."

> "Suppose we could expel sin by this means; look how much we thus expel of sin, so much we expel of virtue: for the matter of them both is the same; remove that, and ye remove them both alike. This justifies the high providence of God, who, though he commands us temperance, justice, continence, yet pours out before us, even to a profuseness, all desirable things, and gives us minds that can wander beyond all limit and satiety. Why should we then affect a rigor contrary to the manner of God and of nature, by abridging or scanting those means, which books freely permitted are, both to the trial of virtue, and the exercise of truth? It would be better done, to learn that the law must needs be frivolous, which goes to restrain things, uncertainly and yet equally working to good and to evil."

Next I quote a passage about Galileo, since the name of Galileo has often been invoked in the debate over recombinant DNA. This passage shows that the connection between the silencing of Galileo and the general decline of intellectual life in seventeenth-century Italy was not invented by the molecular biologists of today but was also obvious to a contemporary eyewitness.

"And lest some should persuade ye, Lords and Commons, that these arguments of learned men's discouragement at this your order are mere flourishes, and not real, I could recount what I have seen and heard in other countries, where this kind of inquisition tyrannizes; when I have sat among their learned men, for that honor I had, and been counted happy to be born in such a place of philosophic freedom, as they supposed England was, while themselves did nothing but bemoan the servile condition into which learning amongst them was brought; that this was it which had damped the glory of Italian wits; that nothing had been there written now these many years but flattery and fustian. There it was that I found and visited the famous Galileo, grown old, a prisoner to the Inquisition, for thinking in astronomy otherwise than the Franciscan and Dominican licencers thought."

My last quotation expresses Milton's patriotic pride in the intellectual vitality of seventeenth-century England, a pride that twentieth-century Americans have good reason to share.

"Lords and Commoners of England, consider what nation it is whereof ye are, and whereof ye are the governors; a nation not slow and dull, but of a quick, ingenious and piercing spirit, acute to invent, subtle and sinewy to discourse, not beneath the reach of any point the highest that human capacity can soar to. Nor is it for nothing that the grave and frugal Transylvanian sends out yearly from the mountainous borders of Russia, and beyond the Hercynian wilderness, not their youth, but their staid men, to learn our language and our theologic arts."

Perhaps, after all, as we struggle to deal with the enduring problems of reconciling individual freedom with public safety, the wisdom of a great poet may be a surer guide than the calculations of risk-benefit analysis.

III. POINTS BEYOND

And indeed there will be time
To wonder, "Do I dare?" and, "Do I dare?"
Time to turn back and descend the stair,
With a bald spot in the middle of my hair. . .
       Do I dare
       Disturb the universe?

T. S. ELIOT, *"The Love-Song of J. Alfred
Prufrock,"* 1917

17

A Distant Mirror

In the spring of 1966 Stanley Kubrick was directing the production of his film *2001, A Space Odyssey* at the MGM studios north of London. He invited me to spend a day at the studios. I arrived early and went in search of Kubrick's building, picking my way among sheds full of used scenery, the refuse of a hundred films. Between the sheds were patches of lush spring grass with sheep contentedly grazing. When I finally found Kubrick I asked him what he was doing with the sheep. "Oh, I don't use them," he said, "but they come in handy when someone wants to do a pastoral scene. Also we have a cafeteria." Sure enough, when we went to lunch at the cafeteria, it was roast lamb.

Kubrick spent the whole morning arranging and rearranging his lights and cameras. His studio was a large empty warehouse. The set was a metal-and-plywood structure which represented a sector of the circular gallery containing the control console of the spaceship "Discovery." The structure creaked and rattled as it moved slowly back and forth in the cradle that supported it. The idea was that the actors would walk inside the structure as it moved, so that they would always be at the lowest point, where the floor was horizontal. The cameras were attached to the structure and moved with it. In this way Kubrick achieved the illusion of people walking around the inside of a revolving ship with centrifugal force giving them an artificial gravity. Wherever they happened to be in the gallery, their local gravity would be pointing straight outward, away from the axis of the ship. This trick could only work for one piece of the gallery at a time. It would not be possible to show two actors in different parts of the gallery simultaneously. But Kubrick was pleased with the way the takes of the gallery were shaping up. He said it was easy to cut quickly from one actor to another instead of showing them together. "I'll bet you nobody in the audience will notice," he said.

That day, only one actor was on the set. His name was Keir Dullea, and he had become famous by playing David in *David and Lisa*. David was a psychotic youth who recoiled in horror from any physical contact with another human being. Keir Dullea had played the role brilliantly. It came as quite a surprise when he walked up to me and shook my hand. He had the lead role in *2001* as the astronaut Bowman. He complained bitterly that Kubrick gave him nothing to do. He had originally accepted the role because he wanted to escape from being typecast as a psychotic youth for the rest of his life. But after three months on the *2001* set he was bored and frustrated. I watched him perform. He walked slowly along the gallery structure as it revolved under him; then when the movement stopped he turned to the control console and pressed some knobs. That was all. The action lasted about a minute. Then Kubrick spent twenty minutes rearranging the lights and cameras. Then Keir climbed back into the gallery and went through his motions again. Then another twenty minutes of standing around while Kubrick fiddled with the lights. Then another one-minute take. And so on. "For Christ's sake, why doesn't he let me act?" said Keir.

I tried to draw Kubrick out by asking him questions about the theme and the characters of his film. He seemed totally uninterested. The only thing he would talk about was gadgetry. He described with great enthusiasm the various tricks he was using to make small models of a spaceship look big. He was inordinately proud of his revolving gallery. He instructed me in the fine points of lighting and camera work. I began to feel as frustrated as Keir Dullea.

To me, the special effects and the technical tricks of film making were only of minor interest. I was amazed that Kubrick should be wasting so much time on these trivialities. I admired Kubrick as the creator of *Doctor Strangelove,* that wonderfully profound and funny story of a nuclear holocaust. The greatness of *Strangelove* lay precisely in the fact that Kubrick took the unthinkable theme of a nuclear holocaust and made it real by showing on the screen the human beings who hold the fate of the world in their hands. The characters in *Strangelove* are real people. A friend of mine who once flew in a training flight of a nuclear B-52 bomber told me that the crew looked and talked exactly like the crew of Kúbrick's "Leper Colony." In *Strangelove* Kubrick had a simple message, and he made his message convincing by his mastery of dialogue and characterization. The dialogue bites; the characters are unforgettable. The absurd story mirrors the absurdity of the real world in which we are living. So I had come to Kubrick's studio expecting to find him at work on another *Strangelove*. Instead I saw only gadgetry. So far as I could see, *2001* had no message, no dialogue and no characters. I complained to Kubrick and asked him why he had left out of this film all the things that made *Strangelove* great. He said, "You will see why when you see the film." Nothing more.

After our roast lamb we went to another building, where there was a big computer. This was not HAL, the soulful computer who is the liveliest character in *2001*. It was a computer of 1960 vintage, busily calculating and printing out pay checks for the MGM employees. Kubrick had had the idea, one of his few hopelessly bad ideas, that he would begin *2001* with some interviews with respectable scientists discussing the probability of an encounter with an alien civilization. He thought this talk show at the beginning would make the story of the film more credible. I was one of the scientists whom he had invited to be interviewed on camera. Of course the fact that I was a scientist had to be expressed visually, and for this purpose the computer was required. If the audience should see me standing in front of this impressive computer, they would know that I am a real scientist.

There was only one snag. The computer made so much noise that our interview was inaudible. Three times the sound technicians rearranged the microphones and started the interview afresh. Each time the man with the earphones shook his head. After the fourth abortive attempt, I suggested to Kubrick that we might talk somewhere else, without the computer. "No," he said firmly. "Tell them to turn the damned thing off." So one of the technicians telephoned the head office. After a short conversation he said, "No good. They need the machine to get the pay checks out tomorrow. If they turn it off they will have to pay the crew extra to work overtime." Kubrick said, "How much?" Another conversation with the head office. "A hundred pounds an hour." "Very good; tell them to give us half an hour." Another call to the head office, and the machine subsided into silence, silence that was costing Kubrick sixpence a second.

We finished our interview within the allotted time and went back to the studio for more takes of the revolving gallery. Kubrick continued for the rest of the afternoon to fiddle with his lights and cameras. At the end of the day I said thank you and goodbye. A few months later I received an apologetic note informing me that the film of our interview had been left on the cutting room floor.

I saw the film, minus the interviews, at the New York premiere in 1968. I was still baffled by it. Kubrick had deliberately avoided the clarity and the fast pace that had made *Strangelove* exciting. He had never relented in his determination not to let Keir Dullea act. As it finally emerged, *2001* was slow, and inhuman, and puzzling. At first I did not like it at all. Only after I had seen it through to the end did I begin to understand why Kubrick had wanted to do it this way.

It is interesting to contrast the film *2001* with the book of the same title which was afterward published by Arthur Clarke. Clarke and Kubrick worked together on the script for the film, Clarke alone on the book. The book tells the same story as the film but in a totally different style. The book explains everything. It gives logical motivations for the human characters, for the malfunctioning of the computer HAL, and for the nature of the alien artifacts that the humans discover. It describes clearly what happens at the end of the story. All the loose ends are cleanly tied up. But this is exactly what Kubrick did not want to do. In the film, motivations are only hinted at, the aliens are completely mysterious, the end of the story is a riddle, and the loose ends remain untied. Kubrick deliberately made the story vague and dreamlike, so that as much as possible could be left to the viewer's imagination.

The message of *Strangelove* was that the people who plan and wage nuclear war are creatures like us, sharing our human weaknesses and inanities. To get this message across, Kubrick chose comedy and witty dialogue as the appropriate tools. But the message of *2001* is exactly the opposite. The message of *2001* is that if ever we confront an alien civilization we will find that the aliens are not creatures like us at all. We will find the aliens so alien that almost nothing they do can be comprehended by us in logical terms. For transmitting this message, the tools of *Strangelove* would have been completely inappropriate. If Kubrick had done what I expected, making a space drama in the style of *Strangelove*, the result would perhaps have been a wittier version of *Star Wars*. Perhaps it would have been as popular as *Star Wars* and an equally overwhelming box office success. But a film of that sort could not have expressed what Kubrick wished to express. He wanted to show an alien civilization as totally inhuman, passing beyond the limits of our comprehension. For this purpose he needed a style of film making that was also inhuman, nonverbal, mystical. Like other great artists, he invented a new style when he had a new message. He was not interested in doing the same thing twice. The film *2001* has many flaws, but it remains a masterpiece. In its strange slow way it embodies the greatness of Kubrick's vision, showing mankind dwarfed and humbled in the presence of something that is, in Haldane's words, "not only queerer than we suppose but queerer than we *can* suppose."

In later years, when *2001* has been revived from time to time, I have often wondered whether I should be sorry or glad that my face is not there on the screen in front of the MGM computer, helping to sell Kubrick's message to the public. On the whole, I am glad. Kubrick certainly did not need my help. The odd thing is that he should ever have thought that he did. How did it happen that a respectable scientist like me was at the MGM studio that day among that crowd of illusionists and actors? I ask myself the question which Lewis Carroll once asked himself under similar circumstances:

> Yet what are all such gaieties to me
> Whose thoughts are full of indices and surds?
> x-squared plus seven-x plus fifty-three
> Equals eleven thirds.

The fact is that I am in some respects a peculiar scientist, just as Lewis Carroll was a peculiar mathematician. Kubrick invited me to his studio because he knew that I am unusual among scientists in having a passionate interest in the problems he was trying to explore. He knew that I am obsessed with the future.

I cannot remember how my obsession with the future began. I believe it may have had its roots in my upbringing among the medieval buildings of Winchester. Winchester is a town in love with the past. The past is there, close and tangible. The house that I lived in as a child was three hundred years old, and William of Wykeham's building in which I went to school was nearly six hundred. The people around me were constantly discussing the fine points of our local history, the details of medieval church architecture, or the latest discovery in the archaeological diggings that were all the time in progress. With the impatience of a child, I reacted strongly against all this. Why were these people so stuck in the past? Why were they so excited about some bishop who lived six hundred years ago? I did not want to go back six hundred years into that dull old world that they loved so much. I would much rather go six hundred years forward. So while they talked learnedly of Chaucer and William of Wykeham, I dreamed of spaceships and alien civilizations. Six hundred years, for anybody who grew up in Winchester, is not a long time. I knew that if I could go six hundred years into the future I would see a lot of things more exciting than old churches.

So I became, and have remained, obsessed with the future. The third part of this book is concerned with that obsession. The future is my third home, after England and America. My wanderings there will be the main theme of the chapters that follow.

I am not a practitioner of the pseudo science of futurology, which has recently become almost a professional discipline, attempting to make quantitative predictions of the short-range future by extrapolating trends from the recent past and present. In the long run, qualitative changes always outweigh quantitative ones. Quantitative predictions of economic and social trends are made obsolete by qualitative changes in the rules of the game. Quantitative predictions of technological progress are made obsolete by unpredictable new inventions. I am interested in the long run, the remote future, where quantitative predictions are meaningless. The only certainty in that remote future is that radically new things will be happening. The only way to explore it is to use our imagination. I accepted Kubrick's invitation because I knew that he was, like me, serious about the future. And I knew that he was, even more than I am, willing to follow his imagination wherever it might lead.

Barbara Tuchman has recently published a marvelous book about the fourteenth century, the century of Chaucer and of William of Wykeham. She called her book *A Distant Mirror,* meaning that she is using the history of that distant past as a mirror to reflect the tragic experiences of the twentieth century and to illuminate our present difficulties. The fourteenth century was indeed a tragic century, not unlike our own, although it produced so much of enduring value in poetry and in buildings. William of Wykeham built six major buildings during his lifetime. All six still stand, and all six are still in use for the same purposes for which he built them. Chaucer's poetry, in spite of changes in the pronunciation and vocabulary of English, has not lost its power to move us. Barbara Tuchman's mirror shows us not only a century of massive human suffering and confusion, but also a brave company of human beings reaching out to us across the centuries with deeds and words of good cheer and encouragement.

I am trying to explore the future as Barbara Tuchman explores the past. The future is my distant mirror. Like her, I use my mirror to place in a larger perspective the problems and difficulties of the present. Like her, I see in my mirror great panoramas of suffering and turmoil. But that is not all. I also see, like her, individual human beings who will reach back to us across the centuries and be grateful for our concern, just as we are grateful to Chaucer and William of Wykeham for the heritage which they have left to us.

It was Einstein who gave us a new scientific vision of the universe as a harmonious whole in which past and future have no absolute significance. Einstein learned in March 1955, shortly before his own death, that Michele Besso had died. Besso had shared Einstein's thoughts in the great days of his youth and had remained for more than fifty years Einstein's closest friend. Einstein wrote a letter of condolence to Besso's sister and son in Switzerland. This is how the letter ended:

Now he has departed from this strange world a little ahead of me. That means nothing. People like us, who believe in physics, know that the distinction between past, present and future is only a stubbornly persistent illusion.

Einstein went serenely to his death four weeks later. His discovery of relativity taught us that in physics the division of space-time into past, present and future is an illusion. He also understood that this division is as illusory in human affairs as it is in physics.

Einstein's vision reinforces the lessons I have learned from Barbara Tuchman's distant mirror and from my own. The past and the future are not remote from us. The people of six hundred years back and of six hundred years ahead are people like ourselves. They are our neighbors in this universe. Technology has caused, and will cause, profound changes in style of life and thought, separating us from our neighbors. All the more precious, then, are the bonds of kinship that tie us all together.

18

Thought Experiments

"The scientific worker of the future will more and more resemble the lonely figure of Daedalus as he becomes conscious of his ghastly mission and proud of it." Of all the scientists I have known, the one who came closest in character to Haldane's Daedalus was not a biologist but a mathematician, by name John von Neumann. To those who knew von Neumann only through his outward appearance, rotund and jovial, it may seem ludicrously inappropriate to compare him with Daedalus. But those who knew him personally, this man who consciously and deliberately set mankind moving along the road that led us into the age of computers, will understand that from a psychological point of view Haldane's portrait of him was extraordinarily prophetic.

During the Second World War, von Neumann worked with great enthusiasm as a consultant to Los Alamos on the design of the atomic bomb. But even then he understood that nuclear energy was not the main theme in man's future. In 1946 he happened to meet his old friend Gleb Wataghin, who had spent the war years in Brazil. "Hello, Johnny," said Wataghin. "I suppose you are not interested in mathematics any more. I hear you are now thinking about nothing but bombs." "That is quite wrong," said von Neumann. "I am thinking about something much more important than bombs. I am thinking about computers."

In September 1948 von Neumann gave a lecture entitled "The General and Logical Theory of Automata," which is reprinted in Volume 5 of his collected works. The lecture is still fresh and readable. Because he spoke in general terms, there is very little in it that is dated. Von Neumann's automata are a conceptual generalization of the electronic computers whose revolutionary implications he was the first to see. An automaton is any piece of machinery whose behavior can be precisely defined in strict mathematical terms. Von Neumann's concern was to lay foundations for a theory of the design and functioning of such machines, which would be applicable to machines far more complex and sophisticated than any we have yet built. He believed that from this theory we could learn not only how to build more capable machines, but also how to understand better the design and functioning of living organisms.

Von Neumann did not live long enough to bring his theory of automata into existence. He did live long enough to see his insight into the functioning of living organisms brilliantly confirmed by the biologists. The main theme of his 1948 lecture is an abstract analysis of the structure of an automaton which is of sufficient complexity to have the power of reproducing itself. He shows that a self-reproducing automaton must have four separate components with the following functions. Component A is an automatic factory, an automaton which collects raw materials and processes them into an output specified by a written instruction which must be supplied from the outside. Component B is a duplicator, an automaton which takes a written instruction and copies it. Component C is a controller, an automaton which is hooked up to both A and B. When C is given an instruction, it first passes the instruction to B for duplication, then passes it to A for action, and finally supplies the copied instruction to the output of A while keeping the original for itself. Component D is a written instruction containing the complete specifications which cause A to manufacture the combined system, A plus B plus C. Von Neumann's analysis showed that a structure of this kind was logically necessary and sufficient for a self-reproducing automaton, and he conjectured that it must also exist in living cells. Five years later Crick and Watson discovered the structure of DNA, and now every child learns in high school the biological identification of von Neumann's four components. D is the genetic materials, RNA and DNA; A is the ribosomes; B is the enzymes RNA and DNA polymerase; and C is the repressor and derepressor control molecules and other items whose functioning is still imperfectly understood. So far as we know, the basic design of every microorganism larger than a virus is precisely as von Neumann said it should be. Viruses are not self-reproducing in von Neumann's sense since they borrow the ribosomes from the cells which they invade.

Von Neumann's first main conclusion was that self-reproducing automata with these characteristics can in principle be built. His second main conclusion, derived from the work of the mathematician Turing, is less well known and goes deeper into the heart of the problem of automation. He showed that there exists in theory a universal automaton, that is to say a machine of a certain definite size and complication, which, if you give it the correct written instruction, will do anything that any other machine can do. So beyond a certain point, you don't need to make your machine any bigger or more complicated to get more complicated jobs done. All you need is to give it longer and more elaborate instructions. You can also make the universal automaton self-reproducing by including it within the factory unit (item A) in the self-reproducing system which I already described. Von Neumann believed that the possibility of a universal automaton was ultimately responsible for the possibility of indefinitely continued biological evolution. In evolving from simpler to more complex organisms you do not have to redesign the basic biochemical machinery as you go along. You have only to modify and extend the genetic instructions. Everything we have learned about evolution since 1948 tends to confirm that von Neumann was right.

As we move into the twenty-first century we shall find von Neumann's analysis increasingly relevant to artificial automata as well as to living cells. Also, as we understand more about biology, we shall find the distinction between electronic and biological technology becoming increasingly blurred. So I pose the problem: Suppose we learn how to construct and program a useful and more or less universal self-reproducing automaton. What does this do to us on the intellectual level? What does it do in particular to the principles of economics, or to our ideas about ecology and social organization?

I shall try to answer these questions by means of a series of thought experiments. A thought experiment is an imaginary experiment which is used to illuminate a theoretical idea. It is a device invented by physicists; the purpose is to concoct an imaginary situation in which the logical contradictions or absurdities inherent in some proposed theory are revealed as clearly as possible. As theories become more sophisticated, the thought experiment becomes more and more useful as a tool for weeding out bad theories and for reaching a profound understanding of good ones. When a thought experiment shows that generally accepted ideas are logically self-contradictory, it is called a "paradox." A large part of the progress of physics during this century has resulted from the discovery of paradoxes and their use as a critique of theory. A thought experiment is often more illuminating than a real experiment, besides being a great deal cheaper. The design of thought experiments in physics has become a form of art in which Einstein was the supreme master. A thought experiment is an entirely different thing from a prediction. The situations that I shall describe are not intended as predictions of things that will actually happen. They are idealized models of developments with which we shall have to come to terms intellectually before we can hope to handle them practically.

My first thought experiment is not my own invention. The basic idea of it was published in an article in *Scientific American* twenty years ago by the mathematician Edward Moore. The article was called "Artificial Living Plants." The thought experiment begins with the launching of a flat-bottomed boat from an inconspicuous shipyard belonging to the RUR Company on the northwest coast of Australia. RUR stands for "Rossum's Universal Robots," a company with a long and distinguished history. The boat moves slowly out to sea and out of sight. A month later, somewhere in the Indian Ocean, two boats appear where one was before. The original boat carried a miniature factory with all the necessary equipment, plus a computer program which enables it to construct a complete replica of itself. The replica contains everything that was in the original boat, including the factory and a copy of the computer program. The construction materials are mainly carbon, oxygen, hydrogen and nitrogen, obtained from air and water and converted into high-strength plastics by the energy of sunlight. Metallic parts are mainly constructed of magnesium, which occurs in high abundance in sea water. Other elements, which occur in low abundance, are used more sparingly as required. The boats are called "artificial plants" because they imitate with machines and computers the life-cycle of the microscopic plants which float in the surface layers of the ocean. It is easy to calculate that after one year there will be a thousand boats, after two years a million, after three years a billion, and so on. It is a population explosion running at a rate several hundred times faster than our own.

The RUR Company did not launch this boat with its expensive

cargo just for fun. In addition to the automatic factory, each boat carries a large tank which it gradually fills with fresh water separated by solar energy from the sea. It is also prepared to use rain water as a bonus when available. The RUR Company has established a number of pumping stations at convenient places around the coast of Australia, each equipped with a radio beacon. Any boat with a full cargo of fresh water is programmed to proceed to the nearest pumping station, where it is quickly pumped dry and sent on its way. After three years, when the boats are dispersed over all the earth's oceans, the RUR Company invites all maritime cities in need of pure water to make use of its services. Up and down the coasts of California and Africa and Peru, pumping stations are built and royalties flow into the coffers of the RUR Company. Deserts begin to bloom—but I think we have heard that phrase before, in connection with nuclear energy. Where is the snag this time?

There are two obvious snags in this thought experiment. The first is the economic snag. The RUR boats may provide us with a free supply of pure water, but it still costs money to use it. Just pumping fresh water onto a desert does not create a garden. In most of the desert areas of the world, even an abundance of fresh water will not rapidly produce wealth. To use the water one needs aqueducts, pumps, pipes, houses and farms, skilled farmers and engineers, all the commodities which will still grow with a doubling time measured in decades rather than in months. The second and more basic snag of the RUR project is the ecological snag. The artificial plants have no natural predators. In the third year of its operation, the RUR Company is involved in lawsuits with several shipping companies whose traffic the RUR boats are impeding. In the fifth year, the RUR boats are spread thick over the surface of almost all the earth's oceans. In the sixth year, the coasts of every continent are piled high with wreckage of RUR boats destroyed in ocean storms or in collisions. By this time, it is clear to everybody that the RUR project is an ecological disaster, and further experiments with artificial plants are prohibited by international agreement. But fortunately for me, the prohibition does not extend to thought experiments.

The details of my second thought experiment are partly taken from a story by the science fiction writer Isaac Asimov. We have the planet Mars, a large piece of real estate, completely lacking in economic value because it lacks two essential things, liquid water and warmth. Circling around the planet Saturn is the satellite Enceladus. Enceladus has a mass equal to five percent of the earth's oceans, and a density rather smaller than the density of ice. It is allowable for the purposes of a thought experiment to assume that it is composed of dirty ice and snow, with dirt of a suitable chemical composition to serve as construction material for self-reproducing automata.

The thought experiment begins with a rocket, carrying a small but highly sophisticated payload, launched from the Earth and quietly proceeding on its way to Enceladus. The payload contains an automaton capable of reproducing itself out of the materials available on Enceladus, using as energy source the feeble light of the far-distant sun. The automaton is programmed to produce progeny that are miniature solar sailboats, each carrying a wide, thin sail with which it can navigate in space, using the pressure of sunlight. The sailboats are launched into space from the surface of Enceladus by a simple machine resembling a catapult. Gravity on Enceladus is weak enough so that only a gentle push is needed for the launching. Each sailboat carries into space a small block of ice from Enceladus. The sole purpose of the sailboats is to deliver their cargo of ice safely to Mars. They have a long way to go. First they must use their sails and the weak pressure of sunlight to fight their way uphill against the gravity of Saturn. Once they are free of Saturn, the rest of their way is downhill, sliding down the slope of the Sun's gravity to their rendezvous with Mars.

For some years after the landing of the rocket on Enceladus, the multiplication of automata is invisible from Earth. Then the cloud of little sailboats begins to spiral slowly outward from Enceladus's orbit. As seen from the Earth, Saturn appears to grow a new ring about twice as large as the old rings. After another period of years, the outer edge of the new ring extends far out to a place where the gravitational effects of Saturn and the sun are equal. The sailboats slowly come to a halt there and begin to spill out in a long stream, falling free toward the sun.

A few years later, the nighttime sky of Mars begins to glow bright with an incessant sparkle of small meteors. The infall continues day and night, only more visibly at night. Day and night the sky is warm. Soft warm breezes blow over the land, and slowly warmth penetrates into the frozen ground. A little later, it rains on Mars for the first time in a billion years. It does not take long for oceans to begin to grow.

There is enough ice on Enceladus to keep the Martian climate warm for ten thousand years and to make the Martian deserts bloom. Let us then leave the conclusion of the experiment to the writers of science fiction, and see whether we can learn from it some general principles that are valid in the real world. The result of the experiment is a genuine paradox. The paradox lies in the fact that a finite piece of hardware, which we may build for a modest price once we understand how to do it, produces an infinite payoff, or at least a payoff that is absurdly large by human standards. We seem here to be getting something for nothing, whereas a great deal of hard experience with practical problems has taught us that everything has to be paid for at a stiff price. The paradox forces us to consider the question, whether the development of self-reproducing automata can enable us to override the conventional wisdom of economists and sociologists. I do not know the answer to this question. But I think it is safe to predict that this will be one of the central concerns of human society in the twenty-first century. It is not too soon to begin thinking about it.

Let me illustrate the question with a third thought experiment. One of the by-products of the Enceladus project is a small self-reproducing automaton well adapted to function in terrestrial deserts. It builds itself mainly out of silicon and aluminum which it can extract from ordinary rocks wherever it happens to be. It can extract from the driest desert air sufficient moisture for its internal needs. Its source of energy is sunlight. Its output is electricity, which it produces at moderate efficiency, together with transmission lines to deliver the electricity wherever you happen to need it. There is bitter debate in Congress over licensing this machine to proliferate over our Western states. The progeny of one machine can easily produce ten times the present total power output of the United States, but nobody can claim that it enhances the beauty of the desert landscape. In the end the debate is won by the antipollution lobby. Both of the alternative sources of power, fossil fuels and nuclear energy, are by this time running into severe pollution problems. Quite apart from the chemical and radioactive pollution which they cause, new power plants of both kinds are adding to the burden of waste heat, which becomes increasingly destructive to the environment. In contrast to all this, the rock-eating automaton generates no waste heat at all. It merely uses the energy that would otherwise heat the desert air and converts some of it into useful form. It also creates no smog and no radioactivity. Legislation is finally passed authorizing the automaton to multiply, with the proviso that each machine shall retain a memory of the original landscape at its site, and if for, any reason the site is abandoned the machine is programmed to restore it to its original appearance.

My third thought experiment is again degenerating into fiction, so I will leave it at this point. It appears to avoid the ecological snag that the RUR boats ran into. It raises several new questions that we have to consider. If solar energy is so abundant and so free from problems of pollution, why are we not already using it on a large scale? The answer is simply that capital costs are too high. The self-reproducing automaton seems to be able to side-step the problem of capital. Once you have the prototype machine, the land and the sunshine, the rest comes free. The rock-eater, if it can be made to work at all, overcomes the economic obstacles which hitherto blocked the large-scale use of solar energy.

Does this idea make sense as a practical program for the twenty-first century? One of the unknown quantities which will determine the practicality of such ideas is the generation time of a self-reproducing automaton, the time that it takes on the average for a population of automata to double. If the generation time is twenty years, comparable with a human generation, then the automata do not change dramatically the conditions of human society. In this case they can multiply and produce new wealth only at about the same rate to which we are accustomed in our normal industrial growth. If the generation time is one year, the situation is different. A single machine then produces a progeny of a million in twenty years, a billion in thirty years, and the economic basis of society can be changed in one human generation. If the generation time is a month, the nature of the problem is again drastically altered. We could then cheerfully contemplate demolishing our industries or our cities and rebuilding them in pleasanter ways within a period of a few years.

It is difficult to find a logical basis for guessing what the generation time might be for automata of the kind which I postulated for my three experiments. The only direct evidence comes from biology. We know that bacteria and protozoa, the simplest truly self-reproducing organisms, have generation times of a few hours or days. At the second main level of biological organization, a higher organism such as a bird has a generation time on the order of a year. At the third level of biological organization, represented precariously by the single species *Homo sapiens,* we have a generation time of twenty years. Roughly speaking, we may say that a biochemical automaton can reproduce itself in a day, a higher central nervous system in a year, a cultural tradition in twenty years. With which of these three levels of organization should our artificial automata be compared?

Von Neumann in his 1948 lecture spoke mainly about automata of the logically simplest kind, reproducing themselves by direct duplication. For these automata he postulated a structure appropriate to a single-celled organism. He pictured them as independent units, swimming in a bath of raw materials and paying no attention to one another. This lowest level of organization is adequate for my first experiment but not for my second and third. It is not enough for automata to multiply on Enceladus like bugs on a rotten apple. To produce the effects which I described in the second and third experiments, automata must propagate and differentiate in a controlled way, like cells of a higher organism. The fully developed population of machines must be as well coordinated as the cells of a bird. There must be automata with specialized functions corresponding to muscle, liver and nerve cell. There must be high-quality sense organs, and a central battery of computers performing the functions of a brain.

At the present time the mechanisms of cell differentiation and growth regulation in higher organisms are quite unknown. Perhaps a good way to understand these mechanisms would be to continue von Neumann's abstract analysis of self-reproducing automata, going beyond the unicellular level. We should try to analyze the minimum number of conceptual components which an automaton must contain in order to serve as the germ cell of a higher organism. It must contain the instructions for building every one of its descendants, together with a sophisticated switching system which ensures that descendants of many different kinds multiply and function in a coordinated fashion. I have not seriously tried to carry through such an analysis. Perhaps, now that von Neumann is dead, we shall not be clever enough to complete the analysis by logical reasoning, but will instead have to wait for the experimental embryologists to find out how Nature solved the problem.

My fourth thought experiment is merely a generalized version of the third. After its success with the rock-eating automaton in the United States, the RUR Company places on the market an industrial development kit, designed for the needs of developing countries. For a small down payment, a country can buy an egg machine which will mature within a few years into a complete system of basic industries together with the associated transportation and communication networks. The thing is custom made to suit the specifications of the purchaser. The vendor's guarantee is conditional only on the purchaser's excluding human population from the construction area during the period of growth of the system. After the system is complete, the purchaser is free to interfere with its operation or to modify it as he sees fit.

Another successful venture of the RUR Company is the urban renewal kit. When a city finds itself in bad shape aesthetically or economically, it needs only to assemble a group of architects and town planners to work out a design for its rebuilding. The urban renewal kit will then be programmed to do the job for a fixed fee.

I do not pretend to know what the possibility of such rapid development of industries and reconstruction of cities would do to human values and institutions. On the negative side, the inhuman scale and speed of these operations would still further alienate the majority of the population from the minority which controls the machinery. Urban renewal would remain a hateful thing to people whose homes were displaced by it. On the positive side, the new technology would make most of our present-day economic problems disappear. The majority of the population would not need to concern themselves with the production and distribution of material goods. Most people would be glad to leave economic worries to the computer technicians and would find more amusing ways to spend their time. Again on the positive side, the industrial development kit would rapidly abolish the distinction between developed and developing countries. We would then all alike be living in the postindustrial society.

What would the postindustrial society be like to live in? Haldane in his *Daedalus* tried to describe it:

Synthetic food will substitute the flower-garden and the factory for the dunghill and the slaughterhouse, and make the city at last self-sufficient.

There's many a strong farmer whose heart would break in two
If he could see the townland that we are riding to.
Boughs have their fruit and blossom at all times of the year,
Rivers are running over with red beer and brown beer,
An old man plays the bagpipes in a golden and silver wood,
Queens, their eyes blue like the ice, are dancing in a crowd.

This is a poetic vision, not a sociological analysis. But I doubt whether anybody can yet do better than Haldane did in 1924 in imagining the human aspects of the postindustrial scene.

19

Extraterrestrials

In the year 1918 a brilliant new star, called by astronomers Nova Aquilae, blazed for a few weeks in the equatorial sky. It was the brightest nova of this century. The biologist Haldane was serving with the British Army in India at the time and recorded his observation of the event:

Three Europeans in India looking at a great new star in the Milky Way. These were apparently all of the guests at a large dance who were interested in such matters. Amongst those who were at all competent to form views as to the origin of this cosmoclastic explosion, the most popular theory attributed it to a collision between two stars, or a star and a nebula. There seem, however, to be at least two possible alternatives to this hypothesis. Perhaps it was the last judgment of some inhabited world, perhaps a too successful experiment in induced radioactivity on the part of some of the dwellers there. And perhaps also these two hypotheses are one, and what we were watching that evening was the detonation of a world on which too many men came out to look at the stars when they should have been dancing.

A few words are needed to explain Haldane's archaic language. He used the phrase "induced radioactivity" to mean what we now call nuclear energy. He was writing fifteen years before the discovery of fission made nuclear energy accessible to mankind. In 1924 scientifically educated people were aware of the enormous store of energy that is locked up in the nucleus of uranium and released slowly in the process of natural radioactivity. The equation $E = mc^2$ was already well known. But attempts to speed up or slow down natural radioactivity by artificial means had failed totally. The nuclear physicists of that time did not take seriously the idea that "induced radioactivity" might one day place in men's hands the power to release vast quantities of energy for good or evil purposes. Haldane had the advantage of being an outsider, a biologist unfamiliar with the details of nuclear physics. He was willing to go against the opinion of the experts in suggesting "induced radioactivity" as a possible cause of terrestrial or extraterrestrial disasters.

The example of Nova Aquilae raises several questions which we must answer before we can begin a serious search for evidence of intelligent life existing elsewhere in the universe. Where should we look, and how should we recognize the evidence when we see it? Nova Aquilae was for several nights the second brightest star in the sky. One had to be either very blind or very busy not to see it. Perhaps it was an artifact of a technological civilization, as Haldane suggested. How can we be sure that it was not? And how can we be sure that we are not now missing equally conspicuous evidence of extraterrestrial intelligence through not understanding what we see? There are many strange and poorly understood objects in the sky. If one of them happens to be artificial, it might stare us in the face for decades and still not be recognized for what it is.

In 1959 the physicists Cocconi and Morrison proposed a simple solution to the problem of recognition of artificial objects. They proposed that we listen for radio messages from extraterrestrial civilizations. If indeed such messages are being transmitted by our neighbors in space with the purpose of attracting our attention, then the messages will be coded in a form which makes their artificiality obvious. Cocconi and Morrison solve the recognition problem by assuming that the beings who transmit the message cooperate with us in making it easy to recognize. The message by its very existence proves that its source must be artificial. A year after Cocconi and Morrison made their proposal, Edward Purcell carried their idea a stage further and described an interstellar dialogue of radio signals traveling to and fro across the galaxy:

What can we talk about with our remote friends? We have a lot in common. We have mathematics in common, and physics, and astronomy. . . . So we can open our discourse from common ground before we move into the more exciting exploration of what is not common experience. Of course, the exchange, the conversation, has the peculiar feature of built-in delay. You get your answer back decades later. But you are sure to get it. It gives your children something to live for and look forward to. It is a conversation which is, in the deepest sense, utterly benign. No one can threaten anyone else with objects. We have seen what it takes to send *objects* around, but one can send information for practically nothing. Here one has the ultimate in philosophical discourse—all you can do is exchange ideas, but you can do that to your heart's content.

Founders of religions are not to be held responsible for the dogmas which their followers build upon their words. Cocconi and Morrison merely suggested that we listen for a certain type of message with radio telescopes. Purcell merely expressed in poetic language the joys of discovery and companionship that would be ours if we could achieve a two-way communication with an alien species. Everything that Cocconi and Morrison and Purcell said was true. But in the subsequent twenty years their suggestions have hardened into a dogma. Many of the people who are interested in searching for extraterrestrial intelligence have come to believe in a doctrine which I call the Philosophical Discourse Dogma, maintaining as an article of faith that the universe is filled with societies engaged in long-range philosophical discourse. The Philosophical Discourse Dogma holds the following truths to be self-evident:

1. Life is abundant in the universe.
2. A significant fraction of the planets on which life exists give rise to intelligent species.
3. A significant fraction of intelligent species transmit messages for our enlightenment.

If these statements are accepted, then it makes sense to concentrate our efforts upon the search for radio messages and to ignore other ways of looking for evidence of intelligence in the universe. But to me the Philosophical Discourse Dogma is far from self-evident. There is as yet no evidence either for it or against it. Since it may be true, I am whole-heartedly in favor of searching for radio messages. Since it may be untrue, I am in favor of looking for other evidence of intelligence, and especially for evidence which does not require the cooperation of the beings whose activities we are trying to observe.

In recent years there have been some serious searches for radio messages. The technology of listening has been steadily improved. No messages have yet been detected, but the listeners are not discouraged. Their efforts have so far searched only a tiny fraction of the radio frequencies and directions in which messages might be coming. They have plans for continuing their searches in future with greatly increased efficiency. They do not need to build huge new radio telescopes to scan the sky for messages. All that they need is a modest allotment of time on existing telescopes, and a modest amount of money to build new data-processing receivers which allow a large number of frequencies to be searched in parallel. Several groups of radio astronomers are hoping to implement these plans. I support their efforts and hope they will be successful. If they are successful and actually detect an interstellar message, it will be the greatest scientific discovery of the century, a turning point in human history, a revolution in mankind's view of ourselves and our place in the universe. But unfortunately, to be successful they will need a great deal of luck. They will need political luck to get funds to build their instruments. And they will need scientific luck to get a cooperative alien to send them a message.

If the radio astronomers are unlucky or the aliens unhelpful, no messages will be heard. But the absence of messages does not imply the nonexistence of alien intelligences. It is important to think about other ways of looking for evidence of intelligence, ways which might still work if the Philosophical Discourse Dogma happens to be untrue. We should not tie our searches to any one hypothesis about the nature and motivation of the aliens. The commonwealth of aliens whispering their secrets to one another in a universe abuzz with radio messages is one possibility. Equally possible, perhaps more probable, is a sparsely populated and uncooperative universe, where life is rare, intelligence is very rare, and nobody outside is interested in helping us discover them. Even under these unfavorable conditions the search for intelligence is not hopeless. When we turn aside from radio messages, the problem which Cocconi and Morrison so neatly solved, of learning how to recognize artificial objects as artificial, becomes again the primary concern.

Let us go back to the example of Nova Aquilae. Nobody now takes seriously Haldane's idea that Nova Aquilae was a too successful experiment in nuclear physics. Why not? What has happened since 1924 to make this idea absurd? What happened was not, as one might have expected, that Haldane's alternative suggestion of an accidental collision as the cause of the nova turned out to be correct. In fact, nobody now believes any of the theories which Haldane mentioned. The reason is simply this. As a result of brilliant observational work done during the last twenty years, mostly by Robert Kraft at the Lick Observatory in California, we now know too much about novae to be satisfied with any theory which explains the outburst as some kind of accident. Kraft observed with meticulous care ten faint stars. Each of them is the dim remnant surviving after a nova explosion. One of them is Nova Aquilae. He discovered that certainly seven out of ten, and probably all ten, of these objects have a peculiar structure which sets them apart from the other stars in the sky. Each of them is a double star consisting of one very small hot component and one rarefied cool component. Each of them has the two stars revolving around each other at such a short distance that they are effectively touching. The periods of revolution are all very short. For Nova Aquilae the period is three hours twenty minutes. We still do not understand in detail why double stars of this special type should be associated with nova explosions. One theory is that there is a steady rain of material from the cool component falling onto the surface of the hot component, and this infalling material is cooked to such a high temperature that it occasionally ignites like a hydrogen bomb. This theory may turn out to be right, or it may be superseded by a better theory. In any case, after Kraft's observations we cannot take seriously any theory of the explosions which does not also explain why they occur only in double stars of this special type. All of Haldane's suggestions fail this test. In particular, it is incredible that intelligent beings capable of conducting disastrous experiments in nuclear physics should appear, in many widely separated parts of the sky, always on planets attached to double stars of a rare and peculiar class.

I reject as worthless all attempts to calculate from theoretical principles the frequency of occurrence of intelligent life forms in the universe. Our ignorance of the chemical processes by which life arose on earth makes such calculations meaningless. Depending on the details of the chemistry, life may be abundant in the universe, or it may be rare, or it may not exist at all outside our own planet. Nevertheless, there are good scientific reasons to pursue the search for evidence of intelligence with some hope for a successful outcome.

The essential point which works in our favor as observers is that we are not required to observe the effects of an average intelligent species. It is enough if we can observe the effects of the most spendthrift, the most grandiosely expansionist, or the most technology-mad society in the universe. Unless of course the species excelling all others in these characteristics happens to be our own.

It is easy to imagine a highly intelligent society with no particular interest in technology. It is easy to see around us examples of technology without intelligence. When we look into the universe for signs of artificial activities, it is technology and not intelligence that we must search for. It would be much more rewarding to search directly for intelligence, but technology is the only thing we have any chance of seeing. To decide whether or not we can hope to observe the effects of extraterrestrial technology, we need to answer the following question: What limits does Nature set to the size and scale of activities of an expansionist technological society? The societies whose activities we are most likely to observe are those which have expanded, for whatever good or bad reasons, to the maximum extent permitted by the laws of physics.

Now comes my main point. Given plenty of time, there are few limits to what a technological society can do. Take first the question of colonization. Interstellar distances look forbiddingly large to human colonists, since we think in terms of our short human lifetime. In one man's lifetime we cannot go very far. But a long-lived society will not be limited by a human lifetime. If we assume only a modest speed of travel, say one hundredth of the speed of light, an entire galaxy can be colonized from end to end within ten million years. A speed of one percent of light velocity could be reached by a spaceship with nuclear propulsion, even using our present primitive technology. So the problem of colonization is a problem of biology and not of physics. The colonists may be long-lived creatures in whose sight a thousand years are but as yesterday, or they may have mastered the technique of putting themselves into cold storage for the duration of their voyage. In any case, interstellar distances are no barrier to a species which has millions of years at its disposal. If we assume, as seems to me probable, that advances in physical technology will allow ships to reach one half of light velocity, then intergalactic distances are no barrier either. A society pressing colonization to the limits of the possible will be able to reach and exploit all the resources of a galaxy, and perhaps of many galaxies.

What are the exploitable resources of a galaxy? The raw materials are matter and energy—matter in the form of planets, comets or dust clouds, and energy in the form of starlight. To exploit these resources fully, a technological species must convert the available matter into biological living space and industrial machinery arranged in orbiting shells around the stars so as to utilize all the starlight. There is enough matter in a planet of the size and chemical composition of Jupiter to form an artificial biosphere exploiting fully the light from a star of the size of our sun. In the galaxy as a whole there may not be enough planets to make biospheres around all the stars, but there are other sources of accessible matter which are sufficient for this purpose. For example, the distended envelopes of red-giant stars are accessible to mining operations and provide matter in quantity far more abundant than that contained in planets. The question remains whether it is technically feasible to build the necessary machinery to create artificial biospheres. Given sufficient time, the job can be done. To convince myself that it is feasible, I have made some rough engineering designs of the machinery required to take apart a planet of the size of the earth and to reassemble it into a collection of habitable balloons orbiting around the sun. To avoid misunderstanding, I should emphasize that I do not suggest that we should actually do this to the earth. We shall have enough dead planets to experiment with so that we shall not need to destroy a live one. But in this chapter I am not concerned with what mankind may do in the future. I am only concerned with the observable effects of what other societies may have done in the past. The construction of an artificial biosphere completely utilizing the light of a star is definitely within the capabilities of any long-lived technological species.

Some science fiction writers have wrongly given me the credit for inventing the idea of an artificial biosphere. In fact, I took the idea from Olaf Stapledon, one of their own colleagues:

> Not only was every solar system now surrounded by a gauze of light traps, which focused the escaping solar energy for intelligent use, so that the whole galaxy was dimmed, but many stars that were not suited to be suns were disintegrated, and rifled of their prodigious stores of subatomic energy.

This passage I found in a tattered copy of Stapledon's *Star Maker* which I picked up in Paddington Station in London in 1945.

The Russian astronomer Kardashev has suggested that civilizations in the universe should fall into three distinct types. A type 1 civilization controls the resources of a planet. A type 2 civilization controls the resources of a star. A type 3 civilization controls the resources of a galaxy. We have not yet achieved type 1 status, but we shall probably do so within a few hundred years. The difference in size and power between types 1 and 2, or between types 2 and 3, is a factor of the order of ten billion, unimaginably large by human standards. But the process of exponential economic growth allows this immense gulf to be bridged remarkably rapidly. To grow by a factor of ten billion takes thirty-three doubling times. A society growing at the modest rate of one percent per year will make the transition from type 1 to type 2 in less than 2500 years. The transition from type 2 to type 3 will take longer than this, since it requires interstellar voyages. But the periods of transition are likely to be comparatively brief episodes in the history of any long-lived society. Hence Kardashev concludes that if we ever discover an extraterrestrial civilization, it will probably belong clearly to type 1, 2 or 3 rather than to one of the brief transitional phases.

In the long run, the only limits to the technological growth of a society are internal. A society has always the option of limiting its growth, either by conscious decision or by stagnation or by disinterest. A society in which these internal limits are absent may continue its growth forever. A society which happens to possess a strong expansionist drive will expand its habitat from a single planet (type 1) to a biosphere exploiting an entire star (type 2) within a few thousand years, and from a single star to an entire galaxy (type 3) within a few million years. A species which has once passed beyond type 2 status is invulnerable to extinction by even the worst imaginable natural or artificial catastrophe. When we observe the universe, we have a better chance of discovering a society that has expanded into type 2 or type 3 than one which has limited itself to type 1, even if the expansionist societies are as rare as one in a million.

Having defined the scale of the technological activities we may look for, I finally come to the questions which are of greatest interest to astronomers: What are the observable consequences of such activities? What kinds of observations will give us the best chance of recognizing them if they exist? It is convenient to discuss these questions separately for civilizations of type 1, 2 and 3.

A type 1 civilization is undetectable at interstellar distances except by radio. The only chance of discovering a type 1 civilization is to follow the suggestion of Cocconi and Morrison and listen for radio messages. This is the method of search that our radio astronomers have followed for the last twenty years.

A type 2 civilization may be a powerful radio source or it may not. So long as we are totally ignorant of the life style of its inhabitants, we cannot make any useful estimate of the volume or nature of their radio emissions. But there is one kind of emission which a type 2 civilization cannot avoid making. According to the second law of thermodynamics, a civilization which exploits the total energy output of a star must radiate away a large fraction of this energy in the form of waste heat. The waste heat is emitted into space as infrared radiation, which astronomers on earth can detect. Any type 2 civilization must be an infrared source with power comparable to the luminosity of a normal star. The infrared radiation will be mainly emitted from the warm outer surface of the biosphere in which the civilization lives. The biosphere will presumably be maintained at roughly terrestrial temperatures if creatures containing liquid water are living in it. The heat radiation from its surface then appears mainly in a band of wavelengths around ten microns (about twenty times the wavelength of visible light). The ten-micron band is fortunately a convenient one for infrared astronomers to work with, since our atmosphere is quite transparent to it.

After Cocconi and Morrison had started the scientific discussion of extraterrestrial intelligence, I made the suggestion that astronomers looking for artificial objects in the sky should begin by looking for strong sources of ten-micron infrared radiation. Of course it would be absurd to claim that evidence of intelligence has been found every time a new infrared source is discovered. The argument goes the other way. If an object in the sky is not an infrared source, then it cannot be the home of a type 2 civilization. So I suggested that astronomers should first make a survey of the sky to compile a catalog of infrared sources, and then look carefully at objects in the catalog with optical and radio telescopes. Using these tactics, the search for radio messages would have greatly improved chances of success. Instead of searching for radio messages over the whole sky, the radio astronomer could concentrate his listening upon a comparatively small number of accurately pinpointed directions. If one of the infrared sources turned out to be also a source of peculiar optical or radio signals, then one could begin to consider it a candidate for possible artificiality.

When I made this proposal twenty years ago, infrared astronomy had hardly begun. Only a few pioneers had started to look for infrared sources, using small telescopes and simple detecting equipment. Now the situation is quite different. Infrared astronomy is a major branch of astronomy. The sky has been surveyed and catalogs of sources exist. I do not claim any credit for this. The astronomers who surveyed the sky and compiled the catalogs were not looking for type 2 civilizations. They were just carrying one step further the traditional mission of astronomers, searching the sky to find out what is there.

Up to now, the infrared astronomers have not found any objects that arouse suspicions of artificiality. Instead they have found a wonderful variety of natural objects, some of them within our galaxy and others outside it. Some of the objects are intelligible and others are not. A large number of them are dense clouds of dust, kept warm by hot stars which may or may not be visible. When the hot star is invisible such an object is called a "cocoon star," a star hidden in a cocoon of dust. Cocoon stars are often found in regions of space where brilliant newborn stars are also seen, for example in the great nebula in the constellation Orion. This fact makes it likely that the cocoon is a normal but short-lived phase in the process of birth of a star.

Superficially, there seems to be some similarity between a cocoon star and a type 2 civilization. In both cases we have an invisible star surrounded by a warm opaque shell which radiates strongly in the infrared. Why, then, does nobody believe that type 2 civilizations are living in the cocoon stars that have now been discovered? First, the cocoons are too luminous. Most of them are radiating hundreds or thousands of times as much energy as the sun. Stars with luminosity as high as this are necessarily short-lived by astronomical standards. A type 2 civilization would be much more likely to exist around a long-lived star like the sun. The infrared radiation which it emits would be hundreds of times fainter than the radiation which we detect from most of the cocoons. A second reason for not believing that cocoons are artificial is that their temperatures are too high to be appropriate for biospheres. Most of them have temperatures higher than 300 degrees centigrade, far above the range in which life as we know it can exist. A third reason is that there is direct visual evidence for dense dust clouds in the neighborhood of cocoons. We have no reason to expect that a type 2 civilization would find it necessary to surround itself with a smoke screen. The fourth and most conclusive reason for regarding cocoons as natural objects is the general context in which they occur. One sees in the same region of space new stars being born and large diffuse dust clouds condensing. The cocoons must be causally related to these other natural processes with which they are associated.

I have to admit that in the twenty years since I made my suggestion, infrared astronomy, with all its brilliant successes, has failed to produce evidence of type 2 civilizations. Should we then give up hope of its ever doing so? I do not believe we should. We can expect to find candidates for type 2 civilizations only when we explore infrared sources a hundred times fainter than the spectacular ones which the astronomers have observed so far. An astronomer prefers to spend his time at the telescope studying in detail one conspicuously interesting object, rather than cataloguing a long list of dim sources for future investigation. I do not blame the astronomers for skimming the cream off the bright sources before returning to the tedious work of surveying the faint ones. We will have to wait a few years before we have a complete survey of sources down to the luminosity of the sun. Only when we have a long list of faint sources can we hope that candidates for type 2 civilizations will appear among them. And we shall not know whether to take these candidates seriously until we have learned at least as much about the structure and distribution of the faint sources as we have now found out about the bright ones.

A type 3 civilization in a distant galaxy should produce emissions of radio, light and infrared radiation with an apparent brightness comparable with those of a type 2 civilization in our own galaxy. In particular, a type 3 civilization should be detectable as an extragalactic infrared source. However, a type 3 civilization would be harder than a type 2 to recognize, for two reasons. First, our ideas about the behavior of a type 3 civilization are even vaguer and more unreliable than our ideas about type 2. Second, we know much less about the structure and evolution of galaxies than we do about the birth and death of stars, and consequently we understand the naturally occurring extragalactic infrared sources even more poorly than we understand the natural sources in our galaxy. We understand the cocoon stars at least well enough to be confident that they are not type 2 civilizations. We do not understand the extragalactic infrared sources well enough to be confident of anything. We cannot expect to recognize a type 3 civilization for what it is until we have thoroughly explored the many strange and violent phenomena that we see occurring in the nuclei of distant galaxies.

Is it possible that a type 3 civilization could exist in our own galaxy? This is a question which deserves more serious thought than has been given to it. The answer is negative if we think of a type 3 civilization as overrunning the galaxy with ruthless efficiency and exploiting the light of every available star. However, other kinds of type 3 civilization are conceivable. One attractive possibility is a civilization based on vegetation growing freely in space rather than on massive industrial hardware. A type 3 civilization might use comets rather than planets for its habitat, and trees rather than dynamos for its source of energy. If such a civilization does not already exist, perhaps we shall one day create it ourselves.

But I must leave these idle dreams to a later chapter and come back to the subject of this one. The subject of this chapter is observation. I do not believe we yet know enough about stars, planets, life and mind to give us a firm basis for deciding whether the presence of intelligence in the universe is probable or improbable. Many biologists and chemists have concluded from inadequate evidence that the development of intelligent life should be a frequent occurrence in our galaxy. Having examined their evidence and heard their arguments, I consider it just as likely that no intelligent species other than our own has ever existed. The question can only be answered by observation.

From the discussion of Nova Aquilae, of civilizations of types 1, 2 and 3, and of the infrared sources, I draw the general conclusion that the best way to look for artificial objects in the sky is to look for natural objects in as many different ways as possible. It is not likely that we can guess correctly what an artificial object should look like. Our best chance is to search for a great variety of natural objects and to try to understand them in detail. When we have found an object that defies natural explanation, we may begin to wonder whether it might be artificial. A reasonable long-range program of searching for evidence of intelligence in the universe is indistinguishable from a reasonable long-range program of general astronomical exploration. We should go ahead with the exploration of the cosmos on all available channels, with visible light, radio, infrared, ultraviolet, x-rays, cosmic rays and gravitational waves. Only by observing on many channels simultaneously shall we learn enough about the objects which we find to tell whether they are natural or artificial. And our program of exploration will bring a rich harvest of discoveries of natural objects, whether or not we are lucky enough to find among them any artificial ones.

20

Clades and Clones

When I was seventeen years old I went to Wales in midwinter with a group of students from the Cambridge University Mountaineering Club. We stayed at the Helyg hut near Capel Curig and climbed on the buttresses of Tryfan in swirling mist and rain and occasional snow. In those days nobody thought of wearing a hard hat for rock climbing. If you were third on the rope, you were supposed to watch out for pebbles dislodged by the climbers above you. I didn't watch out, and a small, sharp pebble severed one of the little arteries in my scalp. The cut was tiny but it bled spectacularly. I untied myself and shouted up into the mist that I had had enough of rock climbing and was going home. I walked down to the nearest road, hoping to get a ride into Capel Curig. I was prepared for a long walk, since this was wartime and gasoline was available only to people driving on official business. Very few cars ever came over the mountain roads, and the short December daylight was already fading. To my astonishment, after I had walked for ten minutes down the road, a bus came by and stopped for me. I got in and asked the driver how often he came over the pass. He looked with evident disapproval at my blood-soaked hair and clothes. "Oh, we only run Tuesdays," he said. So I took a one-way ticket to Betws-y-Coed and from there went on down the valley to Llandudno. In Llandudno there is a hospital which has had a lot of experience in patching up damaged rock climbers.

I stayed in the Llandudno hospital for two days. I was put into a ward with nine other patients. I was washed and fed, my hair was cut and my scalp was sewn up. But my efforts to engage the nurses and patients in friendly conversation failed completely. Nobody, except for the doctor who sewed me up, uttered a single word of English in my presence. Everybody else, patients and nurses and visitors, spoke exclusively Welsh and pretended not to understand me when I spoke English. The Welsh language is beautiful and I enjoyed listening to the music in their voices. But their message was unmistakably clear. I was an alien, and the sooner I got onto the train back to England the better.

This was a sobering experience for an English boy accustomed to consider the words "English" and "British" as synonymous. After six hundred years as a conquered people, and seventy years of compulsory education in the language of the conquerors, the Welsh of Llandudno were still Welsh. When one of the oppressors happened to fall helpless into their hands, they tended his wounds and taught him a lesson he would never forget.

In later years I have seen the same treatment skillfully applied by Swiss Germans to High Germans in Zürich, by Romansh Swiss to German Swiss in Pontresina, by Armenians to Russians in Yerevan, by Pueblo Indians to Anglo Americans at the Jemez pueblo in New Mexico. The smaller and the more evanescent the minority, the more precious is their ancient language, the only weapon they have left with which to humble the conqueror's pride and maintain their own identity as a people. In the whole world there are only two thousand people who speak the language of the Jemez pueblo. If you are a Jemez Indian, you probably drive a Chevrolet and go to work in Albuquerque and have to talk English or Spanish all day on the job. When you come home to the pueblo in the evening, it feels good to hear your children talking the Jemez language, even if they only talk about rock music and baseball. You teach your children not to trade their dignity for tourists' dollars. Jemez pueblo is not a tourist attraction. It is not a museum. It is a living community of people who have succeeded better than most conquered peoples in adapting themselves to the ways of the conquerors without surrendering their cultural heritage and their pride. Like the Welsh in Llandudno, they still have their language. So long as their language lives, they possess an inner fortress that the conquerors cannot penetrate.

The Jews who settled in Israel understood better than anybody the power of language as a moving force in human affairs. When I came as a visitor to Israel, the most impressive sight that I saw, more impressive than museums and universities and cities and farms, was a group of nursery school children in a public park in Haifa chattering to each other in Hebrew, a language that was almost dead a hundred years ago. The revival of Hebrew was the master stroke of the Zionist pioneers. It was that achievement which made all their other achievements possible.

It is an amazing quality of human beings, Jews and Gentiles alike, that we have evolved with an inborn capacity, and perhaps also with an inborn need, for rapid change and diversification of language. This is not what one would naïvely expect. Naïvely, one would expect, when an intelligent species evolves the use of language, that there would be only one language. One would expect that the first speaking animals would evolve a fixed structure of words and meanings, as immutable as the genetic code that evolved three billion years earlier. The wise men who wrote the Bible understood that there was a problem here. They created the legend of the tower of Babel to explain why we have so many languages. Obviously they thought, and many people today think, life would be simpler and human relations easier if we all spoke the same language.

It is true that a world with a universal common language would be a simpler world for bureaucrats and administrators to manage. But there is strong evidence, in our own history and prehistory as well as in the history of contemporary primitive societies, to support the hypothesis that plasticity and diversity of languages played an essential role in human evolution. It is not just an inconvenient historical accident that we have a variety of languages. It was nature's way to make it possible for us to evolve rapidly. Rapid evolution of human capacities demanded that social and biological progress go hand in hand. Biological progress came from random genetic fluctuations that could be significant only in small and genetically isolated communities. To keep a small community genetically isolated and to enable it to evolve new social institutions, it was vitally important that the members of the community could be quickly separated from their neighbors by barriers of language. So our emergence as an intelligent species may have depended crucially on the fact that we have this astonishing ability to switch from Proto-Indo-European to Hittite to Hebrew to Latin to English and back to Hebrew within a few generations. It is likely that in the future our survival and our further development will depend in an equally crucial way on the maintenance of cultural and biological diversity. In the future as in the past, we shall be healthier if we speak many languages and are quick to invent new ones as opportunities for cultural differentiation arise. We now have laws for the protection of endangered species. Why do we not have equally strong laws for the protection of endangered languages?

The analogy between species and languages is only one aspect of a deep-lying analogy between the devices used by nature in biological evolution and the devices used by intelligence in cultural evolution. I am well aware that in drawing analogies of this kind I am treading on dangerous ground. The political abuses of social Darwinism have given a bad name to the whole idea of extending biological concepts into the domain of human societies. Yet there is no reason why a prudent fear of political abuses should cause us to deny the existence of analogies between biological and cultural evolution. The analogies that I have in mind are the following: During the immense interval of time that stretched from about three billion to about half a billion years ago, life made the transition from primitive single-celled organisms to many-celled creatures with diverse and complex structures. We do not know in detail how this great transition came about, but we know that three crucial biological inventions were intimately involved in it. The three fundamental inventions, made by life before the evolution of higher organisms began, were death, sex, and speciation. Death, to enable the future to be different from the past. Sex, to enable genetic characteristics to be rapidly mixed and shared. Speciation, the forming of species isolated from each other by genetic barriers, to make possible the evolution of diversity. These three inventions were all required before living creatures could have elbow room to adapt themselves in shape and behavior to fill the rich variety of ecological niches that their growing diversity was itself beginning to offer to them.

Each of the biological inventions has its analog in the evolution of human culture. The analog of death is tragedy. In every human culture, intelligence and imagination have taken the fact of death and made it into a central theme of ritual, drama and poetry. The great cultures have distilled from death the great works of tragic literature. The analog of sex is romance. In every culture, intelligence has turned sex into a thing of mystery and of beauty. Out of sex we have created the great works of dance, romantic tales and lyric poetry. Finally, we have the third and greatest biological invention, speciation. Intelligence has turned speciation also into a new creative principle, the plasticity and diversification of human languages. Just as speciation gave life freedom to experiment with diversity of form and function, the differentiation of languages gave humanity freedom to experiment with diversity of social and cultural traditions. The flexibility of our social institutions grew out of our multiple linguistic heritage. If ever Welshmen stop speaking Welsh or Jemez Indians stop speaking the Jemez language, all humanity will be the poorer, just as all life was the poorer the day men killed the last moa or the last Steller's sea cow.

The analogy between species and languages can perhaps be carried a step further, to include the processes by which new species and new languages are born. There is some evidence that species commonly originate in groups called clades. *Clade* is a Greek word meaning a branch of a tree, in this case the evolutionary tree on which the twigs are individual species. When some climatic or geographical revolution occurs, upsetting the established balance of nature, not just one new species but a whole clade will appear within a geologically short time. A clade of species seems to be the outcome of an episode of rapid multiplication and diversification of small populations expanding into a new or disturbed habitat. Major evolutionary changes occur by the formation of new clades rather than by the modification of established species. All this is remarkably similar to what happened in Europe after the breakup of the Roman Empire. A great civilization, unified by the Latin language, collapsed. In its place appeared the clade of new Latin-derived languages—French, Spanish, Italian, Portuguese, Romanian—each eventually giving rise to a new civilization with literature and traditions of its own. The clade also contained some other languages—Catalan, Provençal and Romansh—which still must struggle for existence against their more powerful brothers. Other, older groups of languages—the Celtic group including Welsh and the Slavic group including Russian—probably originated in multiple births in a similar way. Only in the case of the Romance languages the process of clade formation occurred within historic times and can be verified from written records. The growth and differentiation of the Romance clade was astonishingly rapid. At most twenty generations separate unified Roman Europe from the Europe of well-established local languages.

In biology, a clone is the opposite of a clade. A clade is a group of populations sharing a common origin but exhibiting genetic diversity so wide that they are barred from interbreeding. A clone is a single population in which all individuals are genetically identical. Clades are the stuff of which great leaps forward in evolution are made. Clones are evolutionary dead ends, slow to adapt and slow to evolve. Clades can occur only in organisms that reproduce sexually. Clones in nature are typically asexual.

All this, too, has its analog in the domain of linguistics. A linguistic clone is a monoglot culture, a population with a single language sheltered from alien words and alien thoughts. Its linguistic inheritance, propagated asexually from generation to generation, tends to become gradually impoverished. The process of impoverishment is easy to see in the declining vocabulary of the great writers of English from Shakespeare to Dickens, not to speak of Faulkner and Hemingway. As the centuries go by, words become fewer and masterpieces of literature become rarer. Linguistic rejuvenation requires the analog of sexual reproduction, the mixture of languages and cross-fertilization of vocabularies. The great flowering of English culture followed the sexual union of French with Anglo-Saxon in Norman England. The clade of Romance languages did not spring from Latin alone but from the cross-fertilization of Latin with the languages of the local barbarian tribes as the empire disintegrated. In human culture as in biology, a clone is a dead end, a clade is a promise of immortality.

Are we to be a clade or a clone? This is perhaps the central problem in humanity's future. In other words, how are we to make our social institutions flexible enough to preserve our precious biological and cultural diversity? There are some encouraging signs that our society is growing more flexible than it used to be. Many styles of behavior are now allowed which thirty or forty years ago were forbidden. In many countries where minority languages were once suppressed, they are now tolerated or even encouraged. Thirty-five years after my visit to Llandudno, I stayed at the house of a friend in Cardiff, the capital city of the English conquerors in Wales, and I was happy to see that the children of my Bengali-speaking host were learning Welsh in the Cardiff city schools. Since they were already fluent in English, Bengali and Arabic, they took Welsh in stride, without difficulty. These children were displaying in a spectacular fashion the gift of cultural and linguistic plasticity with which nature has endowed our species. So long as we continue to raise such children, we shall be in no danger of becoming a clone.

Olaf Stapledon wrote in 1930 a book, *Last and First Men,* which is an attempt to imagine a future history of mankind on the broadest scale. One of the themes that he sees as important in man's future is a philosophical attitude which he calls "The Cult of Evanescence." The cult of evanescence is nothing new. It is strong in Homer's *Iliad* and in the apocryphal book *Ecclesiasticus* of the Hebrew Bible. The essence of it is a profound sense of the nobility and beauty of short-lived creatures, a beauty made the more intense by the fact of their evanescence. The cult is made up of joy and grief inextricably mingled. In Stapledon's vision of the future, the cult of evanescence keeps mankind in balance and in contact with the natural world. It holds in check our tendency to unify and homogenize and obliterate nature's diversity with our technology. It holds in check our tendency to unify and homogenize ourselves. It keeps us forever humble before the universe's prodigality.

The cult of evanescence is sung in the poetry of many languages, especially in the poetry of Gerard Hopkins and Dylan Thomas. Hopkins was an Englishman who found his poetic inspiration in Wales:

> All things counter, original, spare, strange;
> Whatever is fickle, freckled (who knows how?)
> With swift, slow; sweet, sour; adazzle, dim;
> He fathers-forth whose beauty is past change:
>   Praise him.

Hopkins was the only one of our English poets who took the trouble to learn Welsh. He borrowed from the classical Welsh poets some of his most striking devices of rhyme and meter, and even wrote some poems in Welsh himself. Unfortunately, my Welsh friends tell me that Hopkins writing in Welsh is not as good a poet as Dylan Thomas, a Welshman, writing in English. We English have taken from the Welsh far more than we shall ever give back. Dylan Thomas's poetry flows with melodies of youth and evanescence, but under the surface melodies a deeper theme can sometimes be heard, the pride of a spirit imprisoned in an alien culture and an alien language:

> Oh as I was young and easy in the mercy of his means,
> Time held me green and dying
> Though I sang in my chains like the sea.

21

The Greening of the Galaxy

My mother was nineteen years old when the South African war began in 1899, and she lived to see the Americans defeated in Vietnam. She often told me that her memories of England during the South African war made it easy for her to understand what the Vietnam war had done to America. The South African war was for England not just a military and political disaster; it was a collapse of a whole system of values. To my mother and her generation, brought up in the tradition of liberal imperialism, the deepest psychological trauma came not from seeing the great British Empire outwitted and outmaneuvered by the two minuscule Boer republics, but from seeing the British Empire starve the Boers into submission by scorching their earth and herding their women and children into concentration camps. Some of my mother's friends were secretly pro-Boer. To be openly pro-Boer required as much courage as to be openly for Ho Chi Minh in the America of 1965. The war divided families and called loyalties into question. It came suddenly, out of a blue sky, at the end of the long summer of Victorian progress and prosperity.

The worst year was 1901. The old queen died in January, and her death symbolized the passing of the comfortable certainties that English people had come to accept during the sixty-three years of her reign. Through 1901 the war dragged on, as ugly and as inconclusive as the war in Vietnam. England came to the end of 1901 and moved into 1902 with the Boers still fighting and their families still dying of dysentery in the concentration camps. Victorian optimism was gone forever. Doom and gloom were in the air.

At that moment, on Friday, January 24, 1902, six years after writing *The Island of Doctor Moreau,* H. G. Wells gave a lecture at the Royal Institution in London with the title "The Discovery of the Future." Now that the shallow optimism of his countrymen had been replaced by an equally shallow despair, Wells decided that the time had come to tell them a story as different from *Doctor Moreau* as it is possible to imagine. This is the way his lecture ended:

Do not misunderstand me when I speak of the greatness of human destiny. If I may speak quite openly to you, I will confess that, considered as a final product, I do not think very much of myself or (saving your presence) my fellow creatures. I do not think I could possibly join in the worship of humanity with any gravity or sincerity. Think of it. Think of the positive facts. There are surely moods for all of us when one can feel Swift's amazement that such a being should deal in pride. There are moods when one can join in the laughter of Democritus; and they would come oftener were not the spectacle of human littleness so abundantly shot with pain. But it is not only with pain that the world is shot—it is shot with promise. Small as our vanity and carnality makes us, there has been a day of still smaller things. It is the long ascent of the past that gives the lie to our despair. We know now that all the blood and passion of our life was represented in the Carboniferous time by something—something, perhaps, cold-blooded and with a clammy skin, that lurked between air and water, and fled before the giant amphibia of those days. For all the folly, blindness and pain of our lives, we have come some way from that. And the distance we have traveled gives us some earnest of the way we have yet to go. . . .

It is possible to believe that all the past is but the beginning of a beginning, and that all that is and has been is but the twilight of the dawn. It is possible to believe that all the human mind has ever accomplished is but the dream before the awakening. We cannot see, there is no need for us to see, what this world will be like when the day has fully come. We are creatures of the twilight. But it is out of our race and lineage that minds will spring, that will reach back to us in our littleness to know us better than we know ourselves, and that will reach forward fearlessly to comprehend this future that defeats our eyes. All this world is heavy with the promise of greater things, and a day will come, one day in the unending succession of days, when beings, beings who are now latent in our thoughts and hidden in our loins, shall stand upon this earth as one stands upon a footstool, and shall laugh and reach out their hands amidst the stars.

Forty-five years later, at the end of a bigger and even more brutal war, the poet Robinson Jeffers succinctly put the case against Wells's vision of the future:

Names foul in the mouthing.
The human race is bound to defile, I've often noticed it,
Whatever they can reach or name, they'd shit on the morning star
   If they could reach. . . .

The awful power that feeds the life of the stars has been tricked down
   Into the common stews and shambles. . . .

A day will come when the earth will scratch herself and smile and rub
   off humanity.

Wells and Jeffers are both right. Humanity is provisional and
contemptible, big with promise and with mischief. Our path into the
future will not be simple and easy. Wells never said it would be. The
fact that men are ugly does not mean that the universe is ugly. Jeffers
never said it was.

In everything we undertake, either on earth or in the sky, we
have a choice of two styles, which I call the gray and the green. The
distinction between gray and green is not sharp. Only at the ex-
tremes of the spectrum can we say without qualification, this is green
and that is gray. The difference between green and gray is better
explained by examples than by definitions. Factories are gray, gardens are green. Physics is gray, biology is green. Plutonium is gray,
horse manure is green. Bureaucracy is gray, pioneer communities
are green. Self-reproducing machines are gray, trees and children
are green. Human technology is gray, God's technology is green.
Clones are gray, clades are green. Army field manuals are gray,
poems are green.

Why should we not say simply, gray is bad, green is good, and find
a quick path to salvation by embracing green technology and ban-
ning everything gray? Because to answer the world's material needs,
technology has to be not only beautiful but also cheap. We delude
ourselves if we think that the ideology of "Green Is Beautiful" will
save us from the necessity of making difficult choices in the future,
any more than other ideologies have saved us from difficult choices
in the past.

Here on earth, solar energy is one of the great human needs.
Every country, rich or poor, is bathed in an abundance of solar
energy, but we have no cheap and widely available technology for
converting this energy into the fuel and electricity that our daily life
requires. To convert sunlight into fuel or electricity is a scientifically

trivial problem. Many different technologies can in principle make the conversion. But all the existing technologies are expensive. We cannot afford to deploy these technologies on a large enough scale to shift a major fraction of our energy consumption away from our rapidly diminishing reserves of natural gas and oil.

Ted Taylor, after he finished his work on nuclear theft and nuclear safeguards, decided to devote the rest of his working life to the problems of solar energy. He has worked out a design for a system of solar ponds that might possibly, if all goes well, turn out to be radically cheaper than any existing solar energy technology. The idea is to dig large ponds enclosed by dikes and covered with transparent plastic air mattresses, so that the water is heated by sunlight and insulated against cooling winds and evaporation. The water stays hot, summer and winter. Its heat energy can be used for domestic heating, or converted into electricity or into energy of chemical fuels by simple heat engines that are commercially available. If everything works according to plan, the whole system will convert the energy of sunlight falling on the ponds into fuel and electricity with an efficiency of about five percent and at a cost competitive with coal and oil.

I am not making any prediction that Ted's scheme will actually work. Innumerable engineering problems, not to speak of economic and legal snags, must be overcome before we can know whether the scheme's theoretical promise is realizable. I make only the hypothetical statement that if it should happen that everything works as we hope, these ponds will turn the energy economy of the world upside down. Countries with abundant sunshine and water, in particular the poor countries of the humid tropics, will in time become as wealthy as the oil-exporting countries are today. And their wealth will be self-sustaining, not based on a finite store of irreplaceable resources.

Fortunately, this economic transformation of the world does not depend on the success of Ted Taylor's plans. It does not matter much whether Ted's particular idea works or not. Ted is only one man with one design for a solar energy system. Around the world there are hundreds of other groups with other ideas and other designs. All we need to transform the world is one cheap and successful system. It does not have to be Ted's. We should only be careful to give all the groups who come forward with ideas a chance to show what they can do. None of them should be discouraged or excluded on ideological grounds.

Ted's technology is gray rather than green, designed for utility rather than beauty. It is interesting to picture what Ted's solar energy system will do to the physical appearance of our planet, if it should happen that it achieves economic success and is developed on a large scale. We may imagine, as an extreme and unlikely contingency, that the whole world might decide to build enough solar ponds to generate all the energy that is now consumed each year, replacing entirely our present consumption of oil, gas, coal and uranium. This would require that we cover with ponds and plastic about one percent of the land area of the planet. This is about equal to the fraction of the area of the United States now covered with paved highways. The capital costs of the entire solar energy system would also be comparable with the cost of an equal area of highways. In other words, to provide a permanently renewable energy supply for the whole world would only require us to duplicate on a worldwide scale the environmental and financial sacrifices that the United States has made for the automobile. The people of the United States considered the costs of the automobile to be acceptable. I do not venture to guess whether they would consider the same costs worth paying again for a clean and inexhaustible supply of energy. It is likely that in many poorer countries, where energy consumption is smaller and alternative sources of supply are unavailable, people would consider Ted's ponds a great bargain. Some people might even prefer plastic ponds to highways. At least you can walk between ponds more easily than you can walk across highways.

So gray technology is not without value and not without promise. It offers a hope of escape from poverty for the tropical countries around the Caribbean Sea and the Indian Ocean. It is possible to imagine it achieving a major shift of United States energy consumption from fossil fuels to solar energy within twenty-five years, roughly the time it took to build our national highway system. It is important for many reasons that this shift be made rapidly, before the world's supply of oil runs out.

But if we look further ahead than twenty-five or fifty years, green technology has an even greater promise. Especially in the area of solar energy, everything that gray technology can do, green technology can ultimately do better. Long ago God invented the tree, a device for converting air, water and sunlight into fuel and other useful chemicals. A tree is more versatile and more economical than any device our gray technology has imagined. The main drawback of trees as solar energy systems is that we do not know how to harvest them without destroying them and damaging the landscape in which they are growing. The process of harvesting is economically ineffi- cient and aesthetically unpleasant. The chemicals that trees naturally produce do not fit easily into the patterns of use and distribution of an oil-based economy.

Imagine a solar energy system based upon green technology, after we have learned to read and write the language of DNA so that we can reprogram the growth and metabolism of a tree. All that is visible above ground is a valley filled with redwood trees, as quiet and shady as the Muir Woods below Mount Tamalpais in California. These trees do not grow as fast as natural redwoods. Instead of mainly synthesizing cellulose, their cells make pure alcohol or octane or whatever other chemical we find convenient. While their sap rises through one set of vessels, the fuel that they synthesize flows down- ward through another set of vessels into their roots. Underground, the roots form a living network of pipelines transporting fuel down the valley. The living pipelines connect at widely separated points to a nonliving pipeline that takes the fuel out of the valley to wherever it is needed. When we have mastered the technology of reprogram- ming trees, we shall be able to grow such plantations wherever there is land that can support natural forests. We can grow fuel from red- woods in California, from maples in New Jersey, from sycamores in Georgia, from pine forests in Canada. Once the plantations are grown, they may be permanent and self-repairing, needing only the normal attentions of a forester to keep them healthy. If we assume that the conversion of sunlight to chemical fuel has an overall effi- ciency of one-half percent, comparable with the efficiency of growth in natural forests, then the entire present energy consumption of the world could be supplied by growing fuel plantations on about ten percent of the land area. In the humid tropics, less land would be needed for the same output of fuel.

Ted Taylor has proposed a plan for building a solar pond system to supply domestic heat, hot water, electricity and air conditioning to a hundred apartments that are used to house the families of the visiting members who come to work at the Institute for Advanced

Study in Princeton. He hopes that he can build such a system for a total cost of about five thousand dollars per family. The existing oil-heating system would be kept on standby so that the institute members will not freeze when the solar ponds run into difficulties. This plan for a hundred-family demonstration is not just a scaled-down pilot-plant experiment. It is a full-scale test of the solar pond system. One of the beauties of Ted's idea is that solar ponds are cost effective at a hundred-family scale. There is no advantage in going to larger centralized units. Even if the whole world were to be fueled by solar ponds, the system would still be decentralized, with individual units of about the size we are hoping to build in Princeton.

We are not at present contemplating any plan to turn our institute woods into a plantation of artificial trees to supply fuel for the institute's needs. That will come much later, if it ever comes at all. Most of us, given the choice, would rather walk among trees than among plastic ponds. But the technology of artificial trees will take a long time to develop. It may take fifty years, or a hundred, or two hundred. It will probably be a difficult and controversial development, with many mistakes, many failures, many experiments that go well at first but then run into obscure and complicated difficulties. To master the genetic programming of a single species will be only the first step. To make artificial trees survive and flourish in the natural environment, the programmer will need to understand their ecological relationships with thousands of other species that live on their leaves and branches or in the soil among their roots. Perhaps the programming and breeding of artificial trees will always remain an art rather than a science. Perhaps the people who grow fuel plantations will need green thumbs in addition to a knowledge of DNA and computer software. That is another of the advantages of green technology. But the need of mankind for solar energy is urgent. We cannot wait a hundred years for it. If plastic ponds can do the job quicker, we must dig our plastic ponds and leave the trees for our grandchildren.

When mankind moves out from earth into space, we carry our problems with us. The utilization of solar energy will remain one of our central problems. In space as on earth, technology must be cheap if it is to be more than a plaything of the rich. In space as on earth, we shall have a choice of technologies, gray and green, and the economic constraints that limit our choice on earth will have their analogs in space.

Our existing technology for using solar energy in space is based on photovoltaic cells made of silicon. These are excellent for powering scientific instruments but far too expensive for ordinary human needs. Solar ponds may be cheap and efficient on earth but are not an appropriate technology for use in space. It happens that the solar system is divided rather sharply into two zones: an inner zone close to the sun, where sunlight is abundant and water scarce; and an outer zone away from the sun, where water is abundant and sunlight scarce. The earth is on the boundary between the two zones and is the only place, so far as we know, where both sunlight and water are abundant. That is presumably the reason why life arose on earth. It is also the reason why solar ponds are more likely to be useful on earth than anywhere else in the solar system.

We should be looking for technologies that will radically transform the economics of going into space. We need to reduce the costs of space operations, not just by factors of five or ten but by factors of a hundred or a thousand, before the large-scale expansion of mankind into the solar system will be possible. It seems likely that the appropriate technologies will be different in the inner and outer zones. The inner zone, with abundant sunlight and little water, must be a zone of gray technology. Great machines and governmental enterprises can flourish best in those regions of the solar system that are inhospitable to man. Self-reproducing automata built of iron, aluminum and silicon have no need of water. They can proliferate on the moon or on Mercury or in the spaces between, carrying out gigantic industrial projects at no risk to the earth's ecology. They will feed upon sunlight and rock, needing no other raw material for their growth. They will build in space free-floating cities for human habitation. They will bring oceans of water from the satellites of the outer planets, where it is to be had in abundance, to the inner zone, where it is needed.

The proliferation of gray technology in the inner zone of the solar system can alleviate in many ways the economic problems of mankind on earth. The resources of matter and sunlight available in the inner zone exceed by many powers of ten the resources available on the earth's surface. Earth may be directly supplied from space with scarce minerals and industrial products, or even with food and fuel. Earth may be treasured and preserved as a residential parkland, or as a wilderness area, while large-scale mining and manufacturing operations are banished to the moon and the asteroids. Emigration of people from earth will not by itself solve earth's population problem. Earth's population problem must be solved on earth, one way or another, whether or not there is emigration. But the possibility of emigration may indirectly help a great deal to make earth's problem tractable. It may be psychologically and politically easier for the people who remain on earth to accept strict limits on the growth of their population if those who feel an irrepressible emotional commitment to the raising of large families have another place to which they can go.

Where will the emigrants go? Gray technology does not provide a satisfactory answer to this question. Gray technology can build colonies in space in the style of O'Neill's "Island One," cans of metal and glass in which people live hygienic and protected lives, insulated from both the wildness of earth and the wildness of space. We will be lucky if the people in these metal-and-glass cans do not come to resemble more and more as time goes on the people of Huxley's Brave New World. Humanity requires a larger and freer habitat. We do not live by bread alone. The fundamental problem of man's future is not economic but spiritual, the problem of diversity. How do we find room for diversity, either on our crowded earth or in the metal-and-glass cans that our existing space technology provides as living space?

Diversity on the social level means preserving a multiplicity of languages and cultures and allowing room for the growth of new ones, in the face of the homogenizing influences of modern communications and mass media. Diversity on the biological level means allowing parents the right to use the technology of genetic manipulation to raise children healthier or longer-lived or more gifted than themselves. The consequence of allowing to parents freedom of genetic diversification would probably be the splitting of mankind into a clade of noninterbreeding species. It is difficult to imagine that any of our existing social institutions would be strong enough to withstand the strains that such a splitting would impose. The strains would be like the strains caused by the diversity of human skin color, only a hundred times worse. So long as mankind remains confined to this planet, the ethic of human brotherhood must prevail over our desire for diversity. Cultural diversity will inexorably diminish, and biological diversity will be too dangerous to be tolerated.

In the long run, the only solution that I see to the problem of diversity is the expansion of mankind into the universe by means of green technology. Green technology pushes us in the right direction, outward from the sun, to the asteroids and the giant planets and beyond, where space is limitless and the frontier forever open. Green technology means that we do not live in cans but adapt our plants and our animals and ourselves to live wild in the universe as we find it. The Mongolian nomads developed a tough skin and a slit-shaped eye to withstand the cold winds of Asia. If some of our grandchildren are born with an even tougher skin and an even narrower eye, they may walk bare-faced in the winds of Mars. The question that will decide our destiny is not whether we shall expand into space. It is: shall we be one species or a million? A million species will not exhaust the ecological niches that are awaiting the arrival of intelligence.

If we are using green technology, our expansion into the universe is not just an expansion of men and machines. It is an expansion of all life, making use of man's brain for her own purposes. When life invades a new habitat, she never moves with a single species. She comes with a variety of species, and as soon as she is established, her species spread and diversify still further. Our spread through the galaxy will follow her ancient pattern.

To make a tree grow on an asteroid in airless space by the light of a distant sun, we need to redesign the skin of its leaves. In every organism the skin is the crucial part which must be delicately tailored to the demands of the environment. This also is not a new idea.

My conversation with the natives:
"Where do you come from?" I asked them. "We migrated from another planet." "How did you happen to come here and live in a vacuum, when your bodies were designed for living in an atmosphere?" "I can't explain how we got here, that is too complicated, but I can tell you that our bodies gradually changed and adapted to life in a vacuum in the same way as your water-animals gradually became land-animals and your land-animals gradually took to flying. On planets, water-animals generally appear first, air-breathing animals later, and vacuum-animals last." "How do you eat?" "We eat and grow like plants, using sunlight." "But I still don't understand. A plant absorbs juices from the ground and gases from the air, and the sunlight only converts these things into living tissue." "You see these green appendages on our bodies, looking like beautiful emerald wings? They are full of chloroplasts like the ones that make your plants green. A few of your animals

have them too. Our wings have a glassy skin that is airtight and watertight but still lets the sunlight through. The sunlight dissociates carbon dioxide that is dissolved in the blood that flows through our wings, and catalyzes a thousand other chemical reactions that supply us with all the substances we need. . . ."

The quotation is from Konstantin Tsiolkovsky's *Dreams of Earth and Sky,* published in Moscow in 1895, seven years before Wells's lecture on the discovery of the future.

We do not yet know what the asteroids are made of. Many of them are extremely dark in color and have optical characteristics resembling those of a kind of meteorite called carbonaceous chondrite. The carbonaceous chondrites are made of stuff rather like terrestrial soil, containing a fair fraction of water and carbon and other chemicals essential to life. It is possible that we shall be lucky and find that the black asteroids are made of carbonacous chondrite material. Certainly there must be some place in the solar system from which the carbonaceous chrondrites come. If it turns out that the black asteroids are the place, then we have millions of little worlds, conveniently accessible from earth, where suitably programmed trees could take root and grow in the soil as they find it. With the trees will come other plants, and animals, and humans, whole ecologies in endless variety, each little world free to experiment and diversify as it sees fit.

Man's gray technology is also à part of nature. It was, and will remain, essential for making the jump from earth into space. The gray technology was nature's trick, invented to enable life to escape from earth. The green technology of genetic manipulation was another trick of nature, invented to enable life to adapt rapidly and purposefully rather than slowly and randomly to her new home, so that she could not only escape from earth but spread and diversify and run loose in the universe. All our skills are a part of nature's plan and are used by her for her own purposes.

Where do we go next after we have passed beyond the asteroids? The satellites of Jupiter and Saturn are rich in ice and organic nutrients. They are cold and far from the sun, but plants can grow on them if we teach the plants to grow like living greenhouses. There is no reason why a plant cannot grow its own greenhouse, just as a turtle or an oyster grows its own shell. Moving out beyond Jupiter and

Saturn, we come to the realm of the comets. It is likely that the space around the solar system is populated by huge numbers of comets, small worlds a few miles in diameter, composed almost entirely of ice and other chemicals essential to life. We see one of these comets only when it happens to suffer a perturbation of its orbit which sends it plunging close to the sun. Roughly one comet per year is captured into the region near the sun, where it eventually evaporates and disintegrates. If we assume that the supply of distant comets is sufficient to sustain this process over the billions of years that the solar system has existed, then the total population of comets loosely attached to the sun must be numbered in the billions. The combined surface area of these comets is then at least a thousand times that of earth. Comets, not planets, may be the major potential habitat of life in the solar system.

It may or may not be true that other stars have as many comets as the sun. We have no evidence one way or the other. If the sun is not exceptional in this regard, then comets pervade our entire galaxy, and the galaxy is a much friendlier place for interstellar travelers than most people imagine. The average distance between habitable islands in the ocean of space will then not be measured in light-years but will be of the order of a light-day or less.

Whether or not the comets provide convenient way stations for the migration of life all over the galaxy, the interstellar distances cannot be a permanent barrier to life's expansion. Once life has learned to encapsulate itself against the cold and the vacuum of space, it can survive interstellar voyages and can seed itself wherever starlight and water and essential nutrients are to be found. Wherever life goes, our descendants will go with it, helping and guiding and adapting. There will be problems for life to solve in adapting itself to planets of various sizes or to interstellar dust clouds. Our descendants will perhaps learn to grow gardens in stellar winds and in supernova remnants. The one thing that our descendants will not be able to do is to stop the expansion of life once it is well started. The power to control the expansion will be for a short time in our hands, but ultimately life will find its own ways to expand with or without our help. The greening of the galaxy will become an irreversible process.

When we are a million species spreading through the galaxy, the question "Can man play God and still stay sane?" will lose some of its terrors. We shall be playing God, but only as local deities and not as lords of the universe. There is safety in numbers. Some of us will become insane, and rule over empires as crazy as Doctor Moreau's island. Some of us will shit on the morning star. There will be conflicts and tragedies. But in the long run, the sane will adapt and survive better than the insane. Nature's pruning of the unfit will limit the spread of insanity among species in the galaxy, as it does among individuals on earth. Sanity is, in its essence, nothing more than the ability to live in harmony with nature's laws.

I have told this story of the greening of the galaxy as if it were our destiny to be nature's first attempt at an intelligent creature. If there are other intelligences already at large in the galaxy, the story will be different. The galaxy will become even richer in variety of life styles and cultures. We must only be careful not to let our wave of expansion overwhelm and disrupt the ecologies of our neighbors. Before our expansion beyond the solar system begins, we must explore the galaxy thoroughly with our telescopes, and we must know enough about our neighbors to come to them as friends rather than as invaders. The universe is large enough to provide ample living space for all of us. But if, as seems equally probable, we are alone in our galaxy and have no intelligent neighbors, earth's life is still large enough in potentialities to fill every nook and cranny of the universe.

The expansion of life over the universe is a beginning, not an end. At the same time as life is extending its habitat quantitatively, it will also be changing and evolving qualitatively into new dimensions of mind and spirit that we cannot imagine. The acquisition of new territory is important, not as an end in itself, but as a means to enable life to experiment with intelligence in a million different forms.

In 1929 the crystallographer Desmond Bernal wrote a little book, *The World, the Flesh and the Devil,* in which he described the expansion of life into space as one of the chief tasks awaiting mankind. Like me, he was baffled when he tried to imagine what would come afterward. His book ends, as every inquiry into the future must end, with a question:

We want the future to be mysterious and full of supernatural power; and yet these very aspirations, so totally removed from the physical world, have built this material civilization and will go on building it into the future so long as there remains any relation between aspiration and action. But can we count on this? Or rather, have we not here the criterion which will decide the direction of human development? We are on the point of being able to see the effects of our actions and their probable consequences in the future; we hold the future still timidly, but perceive it for the first time, as a function of our own action. Having seen it, are we to turn away from something that offends the very nature of our earliest desires, or is the recognition of our new powers sufficient to change those desires into the service of the future which they will have to bring about?

22

Back to Earth

Anybody who pursues a grand design for the expansion of terrestrial life into the universe had better observe carefully the spirit and style of the people who succeed in living in harmony with nature in the wildernesses of earth. The universe is an archipelago, with small islands of habitable ground separated by vast seas of space. The archipelago that extends up the Pacific coasts of Canada and Alaska from Vancouver to Glacier Bay is in some sense a microcosm of the universe. With these thoughts in mind, I kept a journal of a visit that I made in 1975 to the Canadian Pacific islands where my son and his friends are living.

Monday. Left Vancouver at 5:30 to catch the early ferry to Nanaimo, with Ken Brower and my daughter Emily. Ken drove us north along Vancouver Island to Kelsey Bay. Afternoon ferry from Kelsey Bay to Beaver Cove, arriving 7:30. My son George was at Beaver Cove waiting for us. I had not seen him for three years. Words from Hugh Kingsmill's parody of A. E. Housman's "Shropshire Lad" flashed through my head:

> What, still alive at twenty-two,
> A clean upstanding lad like you.

Because the hour was late and the tide running against us, George did not come in his new six-seater kayak. Instead he came by motorboat with his friend Will, who lives on Swanson Island. George had intended to take us to Hanson Island, but Will's boat had engine trouble and so we all stayed overnight at Will's place. This was lucky. We sat up half the night listening to Will's stories.

Will comes from a Dukhobor village and learned the skills of a pioneer from his Russian-speaking parents. He and his wife came to Swanson Island four years ago with two pairs of hands. Now they have a solid and cozy house for themselves, a guesthouse for their friends, a farm with a Caterpillar tractor, two boats and a blacksmith's forge with a large assortment of machine tools.

Will paid for his two square miles of land by felling and selling a minute fraction of the timber that stood on it. Beyond his homestead, the whole island is untouched forest. The homestead is decorated with wood carvings done by his wife, the house with wrought iron fashioned by Will himself.

The conversation turned to one of my favorite subjects, the colonization of space. I remarked to Will that he and his wife are precisely the people we shall need for homesteading the asteroids. He said, "I don't mind where I go, but I need a place where I can look around at the end of a year and see what I have done."

Tuesday. Facing Will's homestead, two miles away across Blackfish Sound, stands Paul's house on Hanson Island. Paul also lives alone on his island, with his wife and his seven-year-old son, Yasha. Paul and Will are as different as any two people could be. Paul is every inch an intellectual. His house is a ramshackle affair, made of bits of wood and glass, stuck together haphazardly. One side is covered only with a plastic sheet and leaks abominably when it rains. At the dry end are some beautiful rugs, books, and a 250-year-old violin.

We arrived at Paul's place in the morning and found George's kayak at anchor. George had spent the last winter building it, copying the design from the Aleut Indians. He said the Aleuts knew better than anyone else how to travel in these waters. The kayak is blue, covered with animal designs in the Indian style. It has three masts and three sails. George took us inland to see the tree from which he cut the planks for the kayak. Each plank is thirty-five feet long, straight and smooth and polished. Half of the tree is still there, enough for another boat of the same size.

In the afternoon we went out with Yasha in the kayak to look for whales. Since there was no wind and George's crew was inexpert with the paddles, he turned on his outboard motor. I was glad to see that he is no purist. George merely remarked that we must choose, either the whales or the motor, but not both. We chose the motor, and saw the whales only from a distance.

At sunset we lay down in the tents which George had prepared for us, on a rocky point overlooking the sea. The evening was still and clear. Soon we could hear the rhythmic breathing of the whales, puff-puff, puff-puff, lulling us to sleep.

Wednesday. It began to rain at midday and continued for about twelve hours. I was glad to taste the life of the pioneers, not only under sunshine and blue skies. George took us out fishing and quickly caught a fifteen-pound red snapper, enough to make a good supper for us all. He spent the afternoon preparing salads and sauces to go with it. The fish itself he baked over Paul's wood-burning stove.

During the afternoon Jim arrived with his girl friend Allison and their seven-month-old baby. Jim is the man who taught George how to build boats. When George was seventeen he worked for a year with Jim building the *D'Sonoqua,* a forty-eight-foot brigantine with living quarters on board for ten people. After she was finished, Jim and George with a group of their friends lived on her for a year, cruising up and down the coast. Then George decided he was old enough to be his own master, and quit.

This was my first meeting with Jim. I had already heard much about him from George's letters, and expected to encounter another strong pioneer type like Will. The reality was different. Jim came up the beach through the pouring rain on crutches. His back is crippled so that he can barely walk. One stormy night last November, he drove the *D'Sonoqua* onto the rocks, close by the Indian village from whose god she takes her name. That night, he says, the god was angry. Allison was with him on board, seven months pregnant. Also with them were two little girls, daughters of Allison. Jim got them all safely to shore, but they lost the ship and everything they possessed on her. Now, nine months later, *D'Sonoqua* is beached not far from Hanson Island, with gaping holes in her bottom, her inside furnishings rotted and wrecked. Jim has not given her up. Every spare minute, he drags himself to work on her and dreams of getting her afloat. He is skipper of the *D'Sonoqua* still. It was pitch dark when Jim and Allison left. I watched them walk slowly down the beach to the boat, in the dark and pouring rain, Jim on his crutches, Allison carrying the baby in her arms. It was like the last act of *King Lear,* when the crazy old king and his faithful daughter, Cordelia, are led away to their doom; and Lear says:

Upon such sacrifices, my Cordelia,
The gods themselves throw incense.

Tragedy is no stranger to these islands.

Thursday. In the morning it was still raining. Emily and I lay comfortably in our tents while George gave an exhibition of his skill as an outdoorsman. In an open fireplace under the pouring rain, using only wet wood from the forest, a knife and a single match, he lit a fire and cooked pancakes for our breakfast.

In the afternoon the sun came out and we went for a longer ride in the kayak. This time there was some wind, and we could try out the sails. She sailed well downwind, but without a keel she could make no headway upwind. George has made a pair of hydrofoils, which will be fixed to her sides as outriggers and will give her enough grip on the water to sail upwind. But it will take him another month to make the outriggers and put the whole thing together. In the meantime, we have been improving our skill with the paddles.

Since Thursday was our last evening on the island, we went to visit with Paul and his family. When it was almost dark the whales began to sing. Paul had put hydrophones in the water and connected them to speakers in his house. The singing began quietly and grew louder and louder as the whales came close to shore. Then the whole household exploded in sudden frenzy. Paul grabbed his flute, rushed out onto a tree trunk overhanging the water, and began playing weird melodies under the stars. Little Yasha ran beside Paul and punctuated his melodies with high-pitched yelps. And louder and louder came the answering chorus of whale voices from the open door of the house. George took Emily out in a small canoe to see the whales from close at hand. They sat in the canoe a short distance from shore and George began to play his flute too. The whales came close to them, stopping about thirty feet away, as if they enjoyed the music but did not wish to upset the canoe. So the concert continued for about half an hour. Afterward we counted the whales swimming back to the open sea, about fifteen in all. They are of the species popularly known as killer whales, but Paul calls them only by their official name, Orca.

Friday. Our last day. It happened to be shortly after new moon, so that the tides were stronger than usual. We woke early to find the sun shining, sat on our rock overlooking the water and watched the morning birds. Kingfishers skimming below our feet, eagles soaring above our heads. Between Hanson and Swanson islands, about a mile from shore, there is a strong tide race. That morning it was fierce, making a white streak on the blue sea. By and by we saw a little black speck move into the white area and heard the distant putt-putt of a motor. George saw more than Emily and I did. He said quietly, "Those people have some nerve, going with an open boat into that kind of water." A few seconds after he spoke, the black speck disappeared and the noise stopped. George at once moved into action. Taking Ken with him, he ran to Paul's motorboat, an unsinkable affair made of rubber, and within two minutes was on his way out. From the shore we could see nothing for the next half hour. I roused Paul and helped him heat up his stove. Then the rubber boat reappeared and we could make out four figures in it. They came ashore and I helped the old man stagger up the beach, his hand in mine as cold as ice. I was thinking then of Dover Sharp. By saving these two, George had made up for the one I failed to save. We wrapped them in blankets and sat them down by the stove.

An old man and a young man, both loggers on strike, had decided to go out with their aluminum boat to dig clams. It was a lovely morning, clear and still. They never imagined that one could capsize on such a morning. Luckily they had had the sense to cling to their capsized boat and not try to swim to shore. But George said they were close to the end when he found them. The old man had not been able to move his arms or legs any more. In that icy water nobody can last long. While they revived, George cooked hot tea and pancakes on the stove. Then he radioed to their families to send a boat to take them home. The old man afterward told me how it had felt. He said he knew his life was over and he was ready to go under. When the rubber boat appeared he thought he was seeing visions. Only when Ken and George hauled him aboard did he believe it was real. In the afternoon he and I chatted again over cups of tea. He turned out to be intelligent and well read, and he asked me many questions about my life and work at Princeton. And I said, "But it seems to me now the best thing I ever did in Princeton was to raise that boy."

Toward evening a big solid tugboat arrived to take the two loggers home. In the meantime George and Ken had rescued their boat and beached it on Swanson Island, taken their motor apart and soaked the insides in fresh water. So the loggers went home with their boat and their motor intact, ready for another day.

It was now time for us also to depart. George took us in the rubber boat to catch the night ferry going south from Beaver Cove. He was apologetic because we went home empty-handed. He had intended to spend the last day with us salmon-fishing, so that we could take with us two big salmon, one for his friends in Vancouver and one for my family in Princeton. I told him, "You don't need to apologize. Today you went fishing for something bigger than salmon." And that was our goodbye.

23

The Argument from Design

Professional scientists today live under a taboo against mixing science and religion. This was not always so. When Thomas Wright, the discoverer of galaxies, announced his discovery in 1750 in his book *An Original Theory or New Hypothesis of the Universe,* he was not afraid to use a theological argument to support an astronomical theory:

Since as the Creation is, so is the Creator also magnified, we may conclude in consequence of an infinity, and an infinite all-active power, that as the visible creation is supposed to be full of siderial systems and planetary worlds, so on, in like similar manner, the endless immensity is an unlimited plenum of creations not unlike the known universe. . . . That this in all probability may be the real case, is in some degree made evident by the many cloudy spots, just perceivable by us, as far without our starry Regions, in which tho' visibly luminous spaces, no one star or particular constituent body can possibly be distinguished; those in all likelihood may be external creation, bordering upon the known one, too remote for even our telescopes to reach.

Thirty-five years later, Wright's speculations were confirmed by William Herschel's precise observations. Wright also computed the number of habitable worlds in our galaxy:

In all together then we may safely reckon 170,000,000, and yet be much within compass, exclusive of the comets which I judge to be by far the most numerous part of the creation.

His statement about the comets is also correct, although he does not tell us how he estimated their number. For him the existence of so many habitable worlds was not just a scientific hypothesis but a cause for moral reflection:

> In this great celestial creation, the catastrophy of a world, such as ours, or even the total dissolution of a system of worlds, may possibly be no more to the great Author of Nature, than the most common accident in life with us, and in all probability such final and general Doomsdays may be as frequent there, as even Birthdays or mortality with us upon the earth. This idea has something so chearful in it, that I own I can never look upon the stars without wondering why the whole world does not become astronomers; and that men endowed with sense and reason should neglect a science they are naturally so much interested in, and so capable of inlarging the understanding, as next to a demonstration must convince them of their immortality, and reconcile them to all those little difficulties incident to human nature, without the least anxiety.

> All this the vast apparent provision in the starry mansions seem to promise: What ought we then not to do, to preserve our natural birthright to it and to merit such inheritance, which alas we think created all to gratify alone a race of vain-glorious gigantic beings, while they are confined to this world, chained like so many atoms to a grain of sand.

There speaks the eighteenth century. Now listen to the twentieth, speaking through the voices of the biologist Jacques Monod: "Any mingling of knowledge with values is unlawful, forbidden," and of the physicist Steven Weinberg: "The more the universe seems comprehensible, the more it also seems pointless."

If Monod and Weinberg are truly speaking for the twentieth century, then I prefer the eighteenth. But in fact Monod and Weinberg, both of them first-rate scientists and leaders of research in their specialties, are expressing a point of view which does not take into account the subtleties and ambiguities of twentieth-century physics. The roots of their philosophical attitudes lie in the nineteenth century, not in the twentieth. The taboo against mixing knowledge with values arose during the nineteenth century out of the great battle between the evolutionary biologists led by Thomas Huxley and the churchmen led by Bishop Wilberforce. Huxley won the battle, but a hundred years later Monod and Weinberg were still fighting the ghost of Bishop Wilberforce.

The nineteenth-century battle revolved around the validity of an old argument for the existence of God, the argument from design. The argument from design says simply that the existence of a watch implies the existence of a watchmaker. Thomas Wright accepted this argument as valid in the astronomical domain. Until the nineteenth century, churchmen and scientists agreed that it was also valid in the domain of biology. The penguin's flipper, the nest-building instinct of the swallow, the eye of the hawk, all declare, like the stars and the planets in Addison's eighteenth-century hymn, "The hand that made us is divine." Then came Darwin and Huxley, claiming that the penguin and the swallow and the hawk could be explained by the process of natural selection operating on random hereditary variations over long periods of time. If Darwin and Huxley were right, the argument from design was demolished. Bishop Wilberforce despised the biologists, regarding them as irresponsible destroyers of faith, and fought them with personal ridicule. In public debate he asked Huxley whether he was descended from a monkey on his grandfather's or on his grandmother's side. The biologists never forgave him and never forgot him. The battle left scars which are still not healed.

Looking back on the battle a century later, we can see that Darwin and Huxley were right. The discovery of the structure and function of DNA has made clear the nature of the hereditary variations upon which natural selection operates. The fact that DNA patterns remain stable for millions of years, but are still occasionally variable, is explained as a consequence of the laws of chemistry and physics. There is no reason why natural selection operating on these patterns, in a species of bird that has acquired a taste for eating fish, should not produce a penguin's flipper. Chance variations, selected by the perpetual struggle to survive, can do the work of the designer. So far as the biologists are concerned, the argument from design is dead. They won their battle. But unfortunately, in the bitterness of their victory over their clerical opponents, they have made the meaninglessness of the universe into a new dogma. Monod states this dogma with his customary sharpness:

The cornerstone of the scientific method is the postulate that nature is objective. In other words, the *systematic* denial that true knowledge can be got at by interpreting phenomena in terms of final causes, that is to say, of purpose.

Here is a definition of the scientific method that would exclude Thomas Wright from science altogether. It would also exclude some of the most lively areas of modern physics and cosmology.

It is easy to understand how some modern molecular biologists have come to accept a narrow definition of scientific knowledge. Their tremendous successes were achieved by reducing the complex behavior of living creatures to the simpler behavior of the molecules out of which the creatures are built. Their whole field of science is based on the reduction of the complex to the simple, reduction of the apparently purposeful movements of an organism to purely mechanical movements of its constituent parts. To the molecular biologist, a cell is a chemical machine, and the protein and nucleic acid molecules that control its behavior are little bits of clockwork, existing in well-defined states and reacting to their environment by changing from one state to another. Every student of molecular biology learns his trade by playing with models built of plastic balls and pegs. These models are an indispensable tool for detailed study of the structure and function of nucleic acids and enzymes. They are, for practical purposes, a useful visualization of the molecules out of which we are built. But from the point of view of a physicist, the models belong to the nineteenth century. Every physicist knows that atoms are not really little hard balls. While the molecular biologists were using these mechanical models to make their spectacular discoveries, physics was moving in a quite different direction.

For the biologists, every step down in size was a step toward increasingly simple and mechanical behavior. A bacterium is more mechanical than a frog, and a DNA molecule is more mechanical than a bacterium. But twentieth-century physics has shown that further reductions in size have an opposite effect. If we divide a DNA molecule into its component atoms, the atoms behave less mechanically than the molecule. If we divide an atom into nucleus and electrons, the electrons are less mechanical than the atom. There is a famous experiment, originally suggested by Einstein, Podolsky and Rosen in 1935 as a thought experiment to illustrate the difficulties of quantum theory, which demonstrates that the notion of an electron existing in an objective state independent of the experimenter is untenable. The experiment has been done in various ways with various kinds of particles, and the results show clearly that the state of a particle has a meaning only when a precise procedure for observing the state is prescribed. Among physicists there are many different philosophical viewpoints, and many different ways of interpreting the role of the observer in the description of subatomic processes.

But all physicists agree with the experimental facts which make it hopeless to look for a description independent of the mode of observation. When we are dealing with things as small as atoms and electrons, the observer or experimenter cannot be excluded from the description of nature. In this domain, Monod's dogma, "The cornerstone of the scientific method is the postulate that nature is objective," turns out to be untrue.

If we deny Monod's postulate, this does not mean that we deny the achievements of molecular biology or support the doctrines of Bishop Wilberforce. We are not saying that chance and the mechanical rearrangement of molecules cannot turn ape into man. We are saying only that if as physicists we try to observe in the finest detail the behavior of a single molecule, the meaning of the words "chance" and "mechanical" will depend upon the way we make our observations. The laws of subatomic physics cannot even be formulated without some reference to the observer. "Chance" cannot be defined except as a measure of the observer's ignorance of the future. The laws leave a place for mind in the description of every molecule.

It is remarkable that mind enters into our awareness of nature on two separate levels. At the highest level, the level of human consciousness, our minds are somehow directly aware of the complicated flow of electrical and chemical patterns in our brains. At the lowest level, the level of single atoms and electrons, the mind of an observer is again involved in the description of events. Between lies the level of molecular biology, where mechanical models are adequate and mind appears to be irrelevant. But I, as a physicist, cannot help suspecting that there is a logical connection between the two ways in which mind appears in my universe. I cannot help thinking that our awareness of our own brains has something to do with the process which we call "observation" in atomic physics. That is to say, I think our consciousness is not just a passive epiphenomenon carried along by the chemical events in our brains, but is an active agent forcing the molecular complexes to make choices between one quantum state and another. In other words, mind is already inherent in every electron, and the processes of human consciousness differ only in degree but not in kind from the processes of choice between quantum states which we call "chance" when they are made by electrons.

Jacques Monod has a word for people who think as I do and for whom he reserves his deepest scorn. He calls us "animists," believers in spirits. "Animism," he says, "established a covenant between nature and man, a profound alliance outside of which seems to stretch only terrifying solitude. Must we break this tie because the postulate of objectivity requires it?" Monod answers yes: "The ancient covenant is in pieces; man knows at last that he is alone in the universe's unfeeling immensity, out of which he emerged only by chance." I answer no. I believe in the covenant. It is true that we emerged in the universe by chance, but the idea of chance is itself only a cover for our ignorance. I do not feel like an alien in this universe. The more I examine the universe and study the details of its architecture, the more evidence I find that the universe in some sense must have known that we were coming.

There are some striking examples in the laws of nuclear physics of numerical accidents that seem to conspire to make the universe habitable. The strength of the attractive nuclear forces is just sufficient to overcome the electrical repulsion between the positive charges in the nuclei of ordinary atoms such as oxygen or iron. But the nuclear forces are not quite strong enough to bind together two protons (hydrogen nuclei) into a bound system which would be called a diproton if it existed. If the nuclear forces had been slightly stronger than they are, the diproton would exist and almost all the hydrogen in the universe would have been combined into diprotons and heavier nuclei. Hydrogen would be a rare element, and stars like the sun, which live for a long time by the slow burning of hydrogen in their cores, could not exist. On the other hand, if the nuclear forces had been substantially weaker than they are, hydrogen could not burn at all and there would be no heavy elements. If, as seems likely, the evolution of life requires a star like the sun, supplying energy at a constant rate for billions of years, then the strength of nuclear forces had to lie within a rather narrow range to make life possible.

A similar but independent numerical accident appears in connection with the weak interaction by which hydrogen actually burns in the sun. The weak interaction is millions of times weaker than the nuclear force. It is just weak enough so that the hydrogen in the sun burns at a slow and steady rate. If the weak interaction were much stronger or much weaker, any forms of life dependent on sunlike stars would again be in difficulties.

The facts of astronomy include some other numerical accidents that work to our advantage. For example, the universe is built on such a scale that the average distance between stars in an average galaxy like ours is about twenty million million miles, an extravagantly large distance by human standards. If a scientist asserts that the stars at these immense distances have a decisive effect on the possibility of human existence, he will be suspected of being a believer in astrology. But it happens to be true that we could not have survived if the average distance between stars were only two million million miles instead of twenty. If the distances had been smaller by a factor of ten, there would have been a high probability that another star, at some time during the four billion years that the earth has existed, would have passed by the sun close enough to disrupt with its gravitational field the orbits of the planets. To destroy life on earth, it would not be necessary to pull the earth out of the solar system. It would be sufficient to pull the earth into a moderately eccentric elliptical orbit.

All the rich diversity of organic chemistry depends on a delicate balance between electrical and quantum-mechanical forces. The balance exists only because the laws of physics include an "exclusion principle" which forbids two electrons to occupy the same state. If the laws were changed so that electrons no longer excluded each other, none of our essential chemistry would survive. There are many other lucky accidents in atomic physics. Without such accidents, water could not exist as a liquid, chains of carbon atoms could not form complex organic molecules, and hydrogen atoms could not form breakable bridges between molecules.

I conclude from the existence of these accidents of physics and astronomy that the universe is an unexpectedly hospitable place for living creatures to make their home in. Being a scientist, trained in the habits of thought and language of the twentieth century rather than the eighteenth, I do not claim that the architecture of the universe proves the existence of God. I claim only that the architecture of the universe is consistent with the hypothesis that mind plays an essential role in its functioning.

We had earlier found two levels on which mind manifests itself in the description of nature. On the level of subatomic physics, the observer is inextricably involved in the definition of the objects of his observations. On the level of direct human experience, we are aware of our own minds, and we find it convenient to believe that other human beings and animals have minds not altogether unlike our own. Now we have found a third level to add to these two. The peculiar harmony between the structure of the universe and the needs of life and intelligence is a third manifestation of the importance of mind in the scheme of things. This is as far as we can go as scientists. We have evidence that mind is important on three levels. We have no evidence for any deeper unifying hypothesis that would tie these three levels together. As individuals, some of us may be willing to go further. Some of us may be willing to entertain the hypothesis that there exists a universal mind or world soul which underlies the manifestations of mind that we observe. If we take this hypothesis seriously, we are, according to Monod's definition, animists. The existence of a world soul is a question that belongs to religion and not to science.

When my mother was past eighty-five, she could no longer walk as she once did. She was restricted to short outings close to her home. Her favorite walk in those years was to a nearby graveyard which commands a fine view of the ancient city of Winchester and the encircling hills. Here I often walked with her and listened to her talk cheerfully of her approaching death. Sometimes, contemplating the stupidities of mankind, she became rather fierce. "When I look at this world now," she said once, "it looks to me like an anthill with too many ants scurrying around. I think perhaps the best thing would be to do away with it altogether." I protested, and she laughed. No, she said, no matter how enraged she was with the ants, she would never be able to do away with the anthill. She found it far too interesting.

Sometimes we talked about the nature of the human soul and about the Cosmic Unity of all souls that I had believed in so firmly when I was fifteen years old. My mother did not like the phrase Cosmic Unity. It was too pretentious. She preferred to call it a world soul. She imagined that she was herself a piece of the world soul that had been given freedom to grow and develop independently so long as she was alive. After death, she expected to merge back into the world soul, losing her personal identity but preserving her memories and her intelligence. Whatever knowledge and wisdom she had acquired during her life would add to the world soul's store of knowledge and wisdom. "But how do you know that the world soul will want you back?" I said. "Perhaps, after all these years, the world soul will find you too tough and indigestible and won't want to merge with you." "Don't worry about that," my mother replied. "It may take a little while, but I'll find my way back. The world soul can do with a bit more brains."

24

Dreams of Earth and Sky

> Moist Pacific winds, condensing upon the rain forest,
> Enshroud descending tongues of ice. . . .
>
> On one beach, abundant firewood; at another, a better sunset,
> Dependable clam beds, or a chance at abalone. . . .
>
> Sixty-mile days, the blowing of the wind;
> For three weeks we leave no footprints,
> Encamped in storm,
> Our path flowing as water. . . .
>
> In fog,
> No need for radar,
> But only the alertness of the senses;
> The echo of hidden rocks,
> The steepness of the swells upon the shallows. . . .

These are bits of a long poem that my son George sent me a few weeks ago, before he went north for the summer. He has worked hard all winter in his workshop in the woods near Vancouver, building six ocean-going canoes. The six boats are now on their way north, exploring the islands and inlets up the coast of Alaska. Eleven adventurers willing to trust their lives to George's handiwork. It will be three or four months before I hear from him again. I do not worry for his safety. Even when he goes alone, I do not worry. This time he carries the responsibility for twelve, and I know he will bring them back alive.

I am half a world away, asleep in my room at the Hotel Dan in Haifa, Israel. The hotel is big and luxurious and full of American tourists. Sitting in the main dining room and listening to the conversation, you feel as if you have never left California. I avoid the tourists as much as possible, but I enjoy the amenities. I have been giving some lectures on physics and astronomy at the Israel Institute of Technology, known in Haifa as the Technion. Today I gave a mathematical talk, for experts only. I am a theoretical astronomer, more at home with pencil and paper than with a telescope. For me, a galaxy is not just a big swarm of stars in the sky; it is a set of differential equations with solutions that behave in ways we don't yet understand. I was talking today about the equations that are supposed to describe the dynamics of a galaxy. There is a mystery here. When we solve the equations on a computer, the solutions show the stars falling into strongly unstable patterns of motion. When we look at real galaxies in the sky, we do not see these patterns. In science a discrepancy of this sort is always an important clue; it means that something essential has been overlooked, something new is waiting to be discovered. In the case of the galaxies, the discrepancy has two possible explanations. Either our mathematics is wrong, or the galaxies are held stable by some huge concentration of matter that is invisible to our telescopes. I was arguing for the second alternative. I believe that the mathematics is right and that the invisible matter must be there. I had a hard time convincing the Israeli experts. They are young and bright and skeptical. They found a number of weak points in the mathematics. In the end we agreed that the question remains open. To resolve it, we need some better mathematical understanding and also some more precise observations of galaxies. The arguments went on all day at the Technion. It was a long, hot day. In the evening it was a relief to retreat into my air-conditioned room at the Hotel Dan. I flopped into bed and am sleeping soundly. When reason sleeps, strange spirits roam. . . .

George is sitting in the back seat of the elegant little two-seater spaceship that he has just finished building. We are trying it out for the first time. He lets me sit in front with the controls. I am not afraid to fly the ship, with him sitting close behind me. He can reach over and grab the stick if I do anything stupid. I press the takeoff switch and we are on our way. We begin moving up a rickety launching ramp which looks like the start of the big roller-coaster at Belmont Park in San Diego. After leaving the ramp we glide up through the inside of a large building. It is an auditorium with many tiers of empty seats. There is a hole in the roof, and in a few seconds we are outside, heading upward into the night. Gradually my eyes grow accustomed to the darkness and I see the universe of stars and galaxies spread out around us. I zoom ahead, diving from galaxy to galaxy and dodging the occasional star that gets in our way. It seems only a short time ago that George was a little boy afraid of the dark and I was sitting by his bed to calm his fears. Now he is the experienced skipper giving the orders and I am the novice pilot trusting my life to his care and skill. I feel safe, sitting in the cockpit and leaving to him the responsibility of deciding where to go. If anything goes wrong, George will take care of it.

"Let's play homing pigeons," says George. "Yes, let's," say I. Homing pigeons is a game that tests a person's knowledge of astronomy. The rules of the game are simple. You jump into some random and unfamiliar part of the universe and you have to find your way home by recognizing astronomical objects that you have seen before or read about in books. The spaceship has a special device built into it for playing this game. You press the jump button and it makes a random jump. George says, "Now jump," and I press the button.

As we jump, the pattern of stars and galaxies around us changes abruptly. Half the sky is suddenly filled with a black cloud of dust. On the side away from the dust cloud I see brilliant galaxies stretching away to infinity. Nothing in the sky is recognizable. I plunge toward the brightest galaxy and see dimly on the other side of it a cluster of newborn stars that looks familiar. Then another dust cloud blows across our bows and the cluster disappears from view. I move rapidly to the next galaxy. Far away, behind endless arches of stars, I glimpse forms that might be familiar constellations. They scatter into unfamiliar patterns as we approach them.

We cruise around the universe for a long time, filled with the glory of these uncounted galaxies. I am lost but not scared. George sits peacefully behind me, silent as usual. I do not need to worry. I amuse myself, thinking of the most conspicuous astronomical objects that I have read about, and calculating the chances of coming close enough to them to recognize them. There is the Coma cluster of galaxies, hundreds of galaxies tightly bunched together with a pair of giant galaxies at the center. There is the brightest visible quasar, 3C273. There is the giant galaxy M87, with its jet of glowing gas and its halo of globular clusters of stars. I flatter myself that I shall be able to find my way around, as soon as we happen to jump within eyeshot of our own little corner of the universe. I may even be able to teach George a thing or two about astronomy. Of course the game would be much easier if we had a radio telescope. Most of these quasars and giant galaxies are more distinctive as radio sources than as visual objects. With just our eyes, the game is going to take a long time. But we are in no hurry. After a while, my eyes grow tired of searching the sky for landmarks. I rest, and let the ship drift slowly among the stars. We drift as quietly as we once drifted in George's canoe in the Pacific on a windless afternoon in August.

An immeasurable silence, an immeasurable gulf of time passes over us. Our game is forgotten. George and I are no longer homing pigeons. Our home is now not only far away but long ago. There will be no going back. We are free spirits, at home anywhere in the universe, wherever we happen to be. We do not need any more to speak to each other. We have left our old home on earth and the barriers of words that used to separate us from each other.

I look out of the cockpit window and survey the ranks of galaxies shining majestically as ever. Then I become aware of a barely perceptible movement. Slow at first as the hands of a clock, the galaxies are moving. Gradually they begin moving faster. After a long time I can see that they are all moving away from us. Away and away they go, until they are dwindling into the distance, streaming out and away like leaves in a storm. It is the expansion of the universe that we are witnessing. George and I are the first human beings to see the universal expansion through to the end. For a long time we watch the galaxies speeding away into the distance, growing smaller and fainter and finally disappearing. We are left alone, silent in our little ship, with nothing around us but infinite blackness. . . .

I am driving with an Israeli friend over the Golan Heights. It is the first quiet hour I have had since I was dreaming of galaxies in the Hotel Dan. Except for us, nothing moves on the heights. The land is deserted. From time to time we drive past ruins of Syrian farms and villages, abandoned in 1967. There was heavy fighting here in 1973. To me it seems that this empty landscape is full of ghosts, ghosts of villagers and farmers who lived here, ghosts of soldiers who died here. My Israeli companion is perhaps thinking similar thoughts. We drive in silence. We understand each other well enough not to break in on each other's meditations.

In the distance is the towering mass of Mount Hermon. It still has patches of snow on it, defying the June sun. It stands at the corner of this disputed territory, with Israelis on one side of it and Syrians on the other, like F6 standing between British Sudoland and Ostnian Sudoland in the Auden-Isherwood play. I wonder if there is a monastery at the foot of Mount Hermon like the monastery in the play. No, it is Mount Sinai that has the monastery. A pity. I will not have time to visit Mount Sinai. I would have liked to go to the monastery and look into the abbot's crystal ball. The abbot in the play says, "All men see reflected there some fragment of their nature and glimpse a knowledge of those forces by whose operation the future is forecast." Perhaps, after all, that is what happened to me in Haifa. Perhaps that dream of the galaxies was my look into the crystal. "That is not supernatural," says the abbot. "Nothing is revealed but what we have hidden from ourselves." We drive slowly along the narrow roads of the Golan, while I am trying to fix in my mind the details of my voyage among the galaxies. What the abbot said is true. A dream shows us hidden connections between things that our waking minds keep in separate compartments.

And still I am not satisfied. Like M.F. after his first look into the crystal, I want to call the abbot back and take a second look. The vision of the universe that I saw in my dream was only one of many possible universes. It was a mindless, mechanical universe. It was the sort of universe that Steven Weinberg had in view when he wrote, "The more the universe seems comprehensible, the more it also seems pointless." George and I were traveling through that universe like tourists, as I am traveling through the Golan, not belonging to it and not influencing it. I do not accept this vision. I do not believe that we are tourists in our universe. I do not believe that the universe is mindless. I believe the vision reflected only one aspect, and not the deepest, of our nature. We are not merely spectators; we are actors in the drama of the universe. I wish I could take another look into the crystal.

As we begin the long descent across the Golan toward the sea of Galilee, I am thinking how appropriate it is that I have come to this land of Israel for my dreaming. This has been for three thousand years a land of seers and prophets. Even the Hotel Dan, with all its

air-conditioned tourists, stands on the same hill where the prophet
Elijah called down fire from heaven to confound the prophets of
Baal. The name of Elijah brings with it memories from my childhood.
Each summer my father used to take his family to the Three Choirs'
Festival for a week of choral music. The festival rotates in a three-
year cycle between the cathedrals of Gloucester, Worcester and Here-
ford. Since my father wrote a new work for the festival each year, we
were given free tickets to all the performances, including the rehearsals. I liked the rehearsals best, because you could never tell what was
going to happen next. Apart from the new works by my father and
other young composers, the staple diet of the festivals was Bach,
Handel, Mendelssohn and Elgar. The works that the choirs sang with
the most genuine gusto were the three old standbys of the English
choral tradition, Handel's *Messiah,* Mendelssohn's *Elijah* and Elgar's
Dream of Gerontius. Mendelssohn wrote the *Elijah* for the Birmingham festival of 1846 and conducted its first performance there. It was
a tremendous success, and has remained ever since a favorite of the
English choirs. Mendelssohn died a year later at the age of thirty-
eight.

The most dramatic and moving passages of the *Elijah* come after
the confrontation with the prophets of Baal is over. After his great
triumph, Elijah is not exultant but depressed. "But he himself went
a day's journey into the wilderness, and came and sat down under a
juniper tree: and he requested for himself that he might die; and said,
it is enough; now, O Lord, take away my life, for I am not better than
my fathers." Then an angel comes to encourage him, and he goes on
for forty days into the wilderness to Mount Horeb. "And he said, Go
forth, and stand upon the mount before the Lord. And behold, the
Lord passed by, and a great and strong wind rent the mountains, and
brake in pieces the rocks before the Lord; but the Lord was not in
the wind: and after the wind an earthquake; but the Lord was not
in the earthquake: and after the earthquake a fire; but the Lord was
not in the fire: and after the fire a still small voice." Mendelssohn's
music and these words from the Old Testament are ringing in my
head as we come down into the Galilee. In that dream in Haifa I have
seen the greatness and the emptiness of the universe. I have seen the
strong wind, and the earthquake, and the fire, but I have not heard
the still small voice. I have seen the galaxies pass before me, but the
Lord was not in the galaxies. At this point my meditations are inter-

rupted. We arrive at the Ein Gev kibbutz on the eastern shore of the Sea of Galilee, looking across the water at the hills where Jesus of Nazareth walked. We sit down by the sea and eat an open-air lunch of good fresh fish. I put Elijah aside and give my attention to the fish.

Two weeks later, after many lectures and much traveling, the crystal ball comes to me a second time. I am again asleep in a hotel, at the end of another exhausting day. This time, I am seeing the universe from a different angle. The still small voice comes to me, as it came to Elijah, unexpectedly. . . .

I am sitting in the kitchen at home in America, having lunch with my wife and children. I am grumbling as usual about the bureaucracy. For years we have been complaining to lower-level officials and there has never been any response. "Why don't you go straight to the top?" says my wife. "If I were you I would just telephone the head office." I pick up the phone and dial the number. This comes as a big surprise to the children. They know how much I hate telephoning, and they like to tease me about it. Usually I will make all kinds of excuses to avoid making a call, especially when it is to somebody I don't know personally. But this time I take the plunge without hesitation. The children sit silent, robbed of their chance to make fun of my telephone phobia. To my amazement, the secretary answers at once in a friendly voice and asks what I want. I say I would like an appointment. She says, "Good. I have put you down for today at five." I say, "May I bring the children?" She says, "Of course." As I put down the phone I realize with a shock that we have only an hour to get ourselves ready.

I ask the children if they want to come. I tell them we are going to talk to God and they had better behave themselves. Only the two younger girls are interested. I am glad not to have the whole crowd on my hands. So we say goodbye to the others quickly, before they have time to change their minds. It is just the three of us. We slip out of the house quietly and walk into town to the office.

The office is a large building. The inside of it looks like a church, but there is no ceiling. When we look up, we see that the building disappears into the distance like an elevator shaft. We hold hands and jump off the ground and go up the shaft. I look at my watch and see that we have only a few minutes left before five o'clock. Luckily, we are going up fast, and it looks as if we shall be in time for our appointment. Just as the watch says five, we arrive at the top of the shaft and walk out into an enormous throne room. The room has whitewashed walls and heavy black oak beams. Facing us at the end of the room is a flight of steps with the throne at the top. The throne is a huge wooden affair with wicker back and sides. I walk slowly toward it, with the two girls following behind. They are a little nervous, and so am I. It seems there is nobody here. I look at my watch again. Probably God did not expect us to be so punctual. We stand at the foot of the steps, waiting for something to happen.

Nothing happens. After a few minutes I decide to climb the steps and have a closer look at the throne. The girls are shy and stay at the bottom. I walk up until my eyes are level with the seat. I see then that the throne is not empty after all. There is a three-month-old baby lying on the seat and smiling at me. I pick him up and show him to the girls. They run up the steps and take turns carrying him. After they give him back to me, I stay with him for a few minutes longer, holding him in my arms without saying a word. In the silence I gradually become aware that the questions I had intended to raise with him have been answered. I put him gently back on his throne and say goodbye. The girls hold my hands and we walk down the steps together.

Bibliographical Notes

These notes are not intended to be complete. I put them here to avoid peppering the text with footnotes.

Preface

The Collected Works of Leo Szilard: Scientific Papers (Vol. I of Szilard papers), ed. B. T. Feld and G. Weiss Szilard (Cambridge, Mass.: M.I.T. Press, 1972), preface, p. xix.

1. The Magic City

E. Nesbit, *The Magic City* (London: MacMillan, 1910).

A number of books have been written about Edith Nesbit and her friends Eleanor Marx and Edward Aveling. See *The Times Educational Supplement* (London), August 15, 1958, p. 1259; Doris L. Moore, *E. Nesbit, a Biography,* rev. ed. (Philadelphia: Chilton Books, 1966).

J. M. Keynes, *Newton, the Man,* Royal Society of London "Newton Tercentenary Celebrations, 15 July 1946" (Cambridge University Press, 1947), pp. 27–34. Keynes gave his talk at Cambridge in 1942 and died in 1946, leaving a manuscript which was read by his brother Geoffrey Keynes at the Royal Society celebrations.

Robert M. Pirsig, *Zen and the Art of Motorcycle Maintenance: An Inquiry into Values* (New York: Morrow, 1974).

T. S. Eliot, *The Cocktail Party* (London: Faber and Faber, 1950), p. 162. The three lines quoted here were cribbed by Eliot from Percy Bysshe Shelley, *Prometheus Unbound,* Act I, lines 192–195.

263

2. The Redemption of Faust

George Dyson, "The Canterbury Pilgrims," vocal score (Oxford University Press, 1930).

H. T. H. Piaggio, *An Elementary Treatise on Differential Equations and Their Applications* (London: G. Bell and Sons, 1920).

E. T. Bell, *Men of Mathematics* (London: Gollancz, 1937), chap. 20.

P. Terentius Afer, *Heautontimorumenos,* l. 78. See *Terence, with an English Translation by John Sargeaunt,* 2 vols. (London: Heinemann, 1953).

3. The Children's Crusade

Wing Commander John C. MacGown died in September 1979. An obituary notice appeared in the *British Medical Journal,* December 8, 1979, p. 1515. In World War I he fought as a pilot, was shot down, and twice escaped from German prison camps. In World War II he survived 47 operations over Germany. Between and after the wars he followed his vocation as a doctor. For this information I am indebted to his widow, Marjorie MacGown.

The official British government history, *The Strategic Air Offensive Against Germany, 1939–1945,* by Sir Charles Webster and Noble Frankland (4 vols.; London: Her Majesty's Stationery Office, 1961), is generally accurate and objective. The first volume includes an excellent account of the prewar political and military decisions that led to the establishment of Bomber Command as an independent strategic force. A more critical view of the bombing campaign is given by Anthony Verrier, *The Bomber Offensive* (rev. ed.; London: Pan Books Ltd., 1974). The official account of the Operational Research Sections, *The Origins and Development of Operational Research in the Royal Air Force* (Air Publication 3368; London: Her Majesty's Stationery Office, 1963), is inaccurate and uncritical.

I am grateful to the Australian War Memorial in Canberra for allowing me to check the width of the escape hatch of Royal Australian Air Force Lancaster 460 G, which was flown to Australia in 1944 after completing 90 operations with Bomber Command. The width of the Halifax hatch is quoted from memory and may be incorrect.

Kurt Vonnegut, *Slaughterhouse-Five; or, The Children's Crusade; A Duty-Dance with Death* (New York: Seymour Lawrence/Delacorte, 1969).

4. The Blood of a Poet

John Drinkwater, *Abraham Lincoln* (New York: Houghton Mifflin, 1919).

Frank Thompson's letters and poems were published in *There Is a Spirit in Europe . . . A Memoir of Frank Thompson,* collected by T. J. Thompson and E. P. Thompson (London: Gollancz, 1947). Frank's death was described by Raina Sharova in the London *News-Chronicle* of March 8, 1945. A less sympathetic view of Frank's activities in Bulgaria is presented by Stowers Johnson, *Agents Extraordinary* (London: Robert Hale, 1975).

Mrs. Larissa Onyshkevych, an expert in Slavonic languages, informs me that in fact the Glagolitic alphabet was invented by St. Cyril. His students afterward introduced the Cyrillic alphabet and named it in his honor.

A. S. Eddington, *New Pathways in Science* (Cambridge University Press, 1935), chap. 8.

The fire bombing of Japan is well described in the U.S. official history, *The Army Air Forces in World War 2, vol. 5, The Pacific—Matterhorn to Nagasaki,"* ed. W. F. Craven and J. L. Cate (University of Chicago Press, 1948–1958).

5. A Scientific Apprenticeship

Oppenheimer's remark about Lomanitz appears on p. 127 of *In the Matter of J. Robert Oppenheimer, U.S.A.E.C. Transcript of Hearing Before Personnel Security Board* (Washington, D.C.: U.S. Government Printing Office, 1954), referred to henceforth as "Oppenheimer Transcript." Oppenheimer's statement that the physicists have known sin first appeared in a lecture, "Physics in the Contemporary World," *Technology Review, 50,* 201 and 231 (1948), reprinted in *Bulletin of the Atomic Scientists, 4,* 65 and 85 (1948). The statement was quoted twice in *Time,* on February 23 and November 8, 1948. The November 8 issue has Oppenheimer's face on the cover.

Martin J. Sherwin, *A World Destroyed: The Atomic Bomb and the Grand Alliance* (New York: Knopf, 1975), describes the political history of the atomic bomb as revealed in the papers of Roosevelt and Churchill which only became available to historians in 1970 and later. Chapter 8 deals with the discussions which led to the use of the bombs at Hiroshima and Nagasaki. In a foreword, Hans Bethe commends Sherwin's historical accuracy and expresses his own view of the moral issues.

6. A Ride to Albuquerque

In 1975, Richard P. Feynman gave a lecture at the University of California, Santa Barbara, with the title "Los Alamos from Below: Reminiscences of 1943–1945." This will be included in a volume of lectures edited by Lawrence Badash at U.C. Santa Barbara. In the meantime it has been published in *Engineering and Science* (alumni magazine of California Institute of Technology) *39,* 11–30 (January 1976). It gives the best existing picture of wartime Los Alamos.

For a technical account of Feynman's view of physics as he expounded it to me in 1948, see R. P. Feynman, *Quantum Electrodynamics* (New York: Benjamin, 1962). This is a set of notes taken from lectures given by Feynman at the California Institute of Technology in 1953. Unfortunately, the more speculative and idiosyncratic aspects of Feynman's thinking did not survive in the notes.

7. The Ascent of F6

T. S. Eliot, *Murder in the Cathedral* (New York: Harcourt, Brace, 1935).

W. H. Auden and Christopher Isherwood, *The Ascent of F6* (London: Faber and Faber, 1936).

Oppenheimer's remark about "technically sweet" weapons is on p. 81 of Oppenheimer transcript.

Philip M. Stern, *The Oppenheimer Case: Security on Trial* (New York: Harper & Row, 1969), gives a complete and well-documented history of the Oppenheimer hearings.

George Herbert's poem appears in his book *The Temple*, published in 1633.

8. Prelude in E-Flat Minor

Edward Teller, "The Work of Many People," *Science, 121,* 267–275 (February 1955). Teller's statement about Oppenheimer is on p. 710 of Oppenheimer transcript. Oppenheimer's two statements about the hydrogen bomb are on pp. 242 and 229 of the transcript.

9. Little Red Schoolhouse

Technical details of the Triga reactor are described in T. B. Taylor, A. W. McReynolds and F. J. Dyson, "Reactor with Prompt Negative Temperature Coefficient and Fuel Element Therefor," U.S. Patent No. 3127325 (March 31, 1964). In *Perspectives in Modern Physics: Essays in Honor of Hans A. Bethe,* ed. R. E. Marshak (New York: Interscience, 1966), appears a chapter (pp. 573–592) by Frederic de Hoffmann, "High Conversion High-Temperature Reactors," describing the theory of the high-temperature graphite reactor.

For a vivid account of the evolution of the modern airplane, see Nevil Shute, *Slide Rule, the Autobiography of an Engineer* (London: Heinemann, 1954).

10. Saturn by 1970

Parts of this chapter appeared in F. J. Dyson, "Menschheit und Weltall," *Physikalische Blätter, 26,* 7–14 (1970), published by the German Physical Society.

Willy Ley, *Rockets, Missiles and Space Travel* (New York: Viking Press, 1951), p. 147.

A brief technical description of the Orion project was published by J. C. Nance, "Nuclear Pulse Propulsion," Proceedings of the 11th Nuclear Science Symposium of the Institute of Electrical and Electronic Engineers (October 1964). For a history of the project, see F. J. Dyson, "Death of a Project," *Science, 149*, 141–144 (1965).

For Tsiolkovsky's suggestion of solar sailing, see *Works on Rocket Technology by K. E. Tsiolkovsky* (publishing house of the Defense Industry, Moscow, 1947), NASA Technical Translation TTF-243 (Washington, D.C., 1965), pp. 208–209. The suggestion appeared in a book, *Investigation of Universal Space by Reactive Devices* (Kaluga, 1926), which was a revised version of articles with the same title written in 1903 and 1911.

L. Friedman et al., "Solar Sailing—The Concept made Realistic," Proceedings of the American Institute of Aeronautics and Astronautics 16th Aerospace Science Meeting (Huntsville, Ala., January 1978), describe the heliogyro vehicle and the Halley's Comet mission, with references to earlier literature.

11. Pilgrims, Saints and Spacemen

This chapter is taken from F. J. Dyson, *Proc. American Phil. Soc., 122*, 63–68 (1978), published by the American Philosophical Society.

William Bradford, *Of Plymouth Plantation, 1620–1647*, ed. Samuel E. Morison (New York: Knopf, 1952).

Brigham Young, *History of the Church of Jesus Christ of Latter-Day Saints, Period II, from the Manuscript History of Brigham Young and other Original Documents*, vol. 7, ed. B. H. Roberts (Salt Lake City: Deseret News, 1960); *Messages of the First Presidency of the Church of Jesus Christ of Latter-Day Saints, 1833–1964*, Vol. 1, ed. James R. Clark (Salt Lake City: Bookcraft, 1965).

Gerard K. O'Neill, *The High Frontier, Human Colonies in Space* (New York: Morrow, 1977).

E. H. Phelps Brown and Sheila V. Hopkins, "Seven Centuries of Building Wages" and "Seven Centuries of the Prices of Consumables, compared with Builders' Wage-Rates," in *Essays in Economic History*, vol. 2, ed. E. M. Carus-Wilson (London: Edward Arnold, 1962).

12. Peacemaking

Parts of this and the following chapter appeared in *SALT: Problems and Prospects*, ed. Morton A. Kaplan (Morristown, N.J.: General Learning Press, 1973), chap. 9.

F. J. Dyson, "The Future Development of Nuclear Weapons," *Foreign Affairs, 38,* 457–464 (1960).

Louis B. Sohn, "Zonal Disarmament and Inspection," *Bulletin of the Atomic Scientists, 18,* No. 7, 4–10 (September 1962).

N. S. Khrushchev, "The Present International Situation and the Foreign Policy of the Soviet Union," Report to USSR Supreme Soviet on December 12, 1962, *Pravda,* December 13, 1962, translation in *Current Digest of the Soviet Press, 14,* Nos. 51 and 52 (1962).

13. The Ethics of Defense

George F. Kennan, "Reflections on Our Present International Situation," (Woodmont, Conn.: Promoting Enduring Peace, Inc., 1959).

Ann H. Cahn, *Eggheads and Warheads: Scientists and the ABM* (Cambridge, Mass.: M.I.T. Press, 1971), examines from the point of view of a historian the involvement of scientists in the political battle over missile defense.

James Eayrs, "Arms Control on the Great Lakes," *Disarmament and Arms Control, 2,* 372–404 (1964), describes the history of the Rush-Bagot Treaty.

14. The Murder of Dover Sharp

Robert J. Lifton, *Death in Life: the Survivors of Hiroshima* (New York: Random House, 1967).

M. Willrich and T. B. Taylor, *Nuclear Theft: Risks and Safeguards* (Cambridge, Mass.: Bollinger Publishing Co., 1974).

John McPhee, *The Curve of Binding Energy* (New York: Farrar, Straus and Giroux, 1974).

15. The Island of Doctor Moreau

H. G. Wells, *The Island of Doctor Moreau* (London: Heinemann, 1896), chap. 12.

J. B. S. Haldane, *Daedalus, or Science and the Future* (London: Kegan Paul, 1924).

Aldous Huxley, *Brave New World* (London: Chatto and Windus, 1932).

Matthew S. Meselson, "Behind the Nixon Policy for Chemical and Biological Warfare," *Bull. Atomic Scientists, 26,* 23–34 (January 1970), and "Gas Warfare and the Geneva Protocol of 1925," *Bull. Atomic Scientists, 28,* 33–37 (February 1972). These articles contain extracts from Meselson's congressional testimony.

"Research with Recombinant DNA" (Washington, D.C.: National Academy of Sciences, 1977) is a record of the proceedings of the public forum on recombinant DNA, held in March 1977, while the controversy was at its height.

The final words of this chapter are taken from a letter of Matthew Meselson dated September 7, 1978.

16. Areopagitica

For the full text of my testimony, see "Science Policy Implications of DNA Recombinant Molecule Research," Hearings Before the Subcommittee on Science, Research and Technology of the Committee on Science and Technology, U.S. House of Representatives, 95th Congress (Washington, D.C.: U.S. Government Printing Office, 1977), pp. 837–858.

17. A Distant Mirror

Jerome B. Agel, *The Making of Kubrick's 2001* (New York: New American Library, 1970). Arthur C. Clarke, *2001, A Space Odyssey* (New York: New American Library, 1968).

Lewis Carroll, *Phantasmagoria and Other Poems* (London: Mac-Millan, 1919, originally published 1869), p. 153.

Barbara W. Tuchman, *A Distant Mirror: The Calamitous 14th Century* (New York: Knopf, 1978).

Albert Einstein, Michele Besso, *Correspondance 1903–1955*, ed. Pierre Speziali (Paris: Hermann, 1972), letter 215, p. 538, my translation.

18. Thought Experiments

Parts of this chapter appeared in F. J. Dyson, "The Next Industrial Revolution," *The Key Reporter, 42,* No. 3, 2–8 (1977), published by the United Chapters of Phi Beta Kappa.

John von Neumann, "The General and Logical Theory of Automata," pp. 288–328 of vol. 5 of *Collected Works,* ed. A. H. Taub (New York: Macmillan, 1961–1963); *Theory of Self-Reproducing Automata,* edited and completed by Arthur W. Burks (Urbana: University of Illinois Press, 1966). Von Neumann died in 1957, leaving unfinished manuscripts on self-reproducing automata which Burks assembled and made into a book.

Edward F. Moore, "Artificial Living Plants," *Scientific American, 195,* No. 4, 118–126 (October 1956).

Isaac Asimov, *The Martian Way and Other Stories* (Garden City, N.Y.: Doubleday, 1955).

The poem by Haldane quoted without acknowledgment is "The Happy Townland" by W. B. Yeats. (I have not been able to find out whether the Haldane poem quoted on p. 171 was also borrowed from somebody else.)

19. Extraterrestrials

The book *Interstellar Communication,* ed. A. G. W. Cameron (New York: Benjamin, 1963), contains the papers mentioned in the text: G. Cocconi and P. Morrison, "Searching for Interstellar Communications" (chap. 15), E. Purcell, "Radioastronomy and Communication Through Space" (chap. 13), F. J. Dyson, "Search for Artificial Stellar Sources of Infrared Radiation" (chap. 11).

A summary of progress up to 1977 and of plans for the future is contained in *The Search for Extraterrestrial Intelligence,* ed. P. Morrison, J. Billingham and J. Wolfe, NASA publication SP-419 (Washington, D.C.: U.S. Government Printing Office, 1977).

R. P. Kraft, "Binary Stars among Cataclysmic Variables, III. Ten Old Novae," *Astrophys. Journal, 139,* 457–475 (1964).

N. S. Kardashev, "Transmission of Information by Extraterrestrial Civilizations," *Soviet Astronomy AJ, 8,* 217–221 (1964).

Engineering aspects of artificial biospheres are discussed in F. J. Dyson, "The Search for Extraterrestrial Technology," pp. 641–655 of *Perspectives in Modern Physics* (see notes on chap. 9).

Olaf Stapledon, *Last and First Men, and Star Maker* (New York: Dover reprint, 1968), p. 380.

20. Clades and Clones

The title of this chapter is taken from S. M. Stanley, "Clades and Clones in Evolution: Why We Have Sex," *Science, 190,* 382–383 (1975).

Olaf Stapledon, *Last and First Men,* pp. 175–179.

Eleanor Ruggles, *Gerard Manley Hopkins, a Life* (London: John Lane, 1947), chap. 6.

Dylan Thomas, *Deaths and Entrances* (London: Dent and Sons, 1946), p. 66.

21. The Greening of the Galaxy

H. G. Wells, "The Discovery of the Future," *Nature, 65,* 326–331 (1902).

Robinson Jeffers, *The Double Axe and Other Poems, Including Eleven Suppressed Poems* (New York: Liveright, 1977), pp. 56–57, 105, 137.

K. E. Tsiolkovsky, *Dreams of Earth and Sky* (Moscow: Goncharov, 1895; edition edited by B. N. Vorob'yeva; Moscow: USSR Academy of Sciences, 1959; my translation), pp. 43–44.

J. D. Bernal, *The World, the Flesh and the Devil: An Enquiry into the Future of the Three Enemies of the Rational Soul,* 2d ed. (Bloomington: Indiana University Press, 1969), pp. 80–81.

A year before Bernal's book appeared, Robert Nast published an even bolder inquiry into the future of life and intelligence, *Kosmische Hypothesen* (Leipzig: A. und R. Huber's Verlag, 1928). Nast's book was never translated or reprinted. I have often wondered whether Olaf Stapledon was acquainted with it.

22. Back to Earth

For Kingsmill's parody of Housman, see Michael Holroyd, *Hugh Kingsmill, a Critical Biography* (London: Unicorn Press, 1964), p. 140. For the identification of the author I am indebted to Mike O'Loughlin.

23. The Argument from Design

Thomas Wright, *An Original Theory or New Hypothesis of the Universe* (1750), facsimile reprint with introduction by M. A. Hoskin (London: MacDonald; New York: American Elsevier, 1971), pp. 76 and 83–84.

Jacques Monod, *Le Hasard et la Nécessité* (Paris: Editions du Seuil, 1970); *Chance and Necessity*, transl. A. Wainhouse (New York: Knopf, 1971), pp. 21, 31, 176, 180.

Steven Weinberg, *The First Three Minutes* (New York: Basic Books, 1977), p. 154.

A. Einstein, B. Podolsky and N. Rosen, "Can Quantum-Mechanical Description of Physical Reality be Considered Complete?" *Physical Review, 47,* 777–780 (1935); reply by N. Bohr with same title, *Physical Review, 48,* 696–702 (1935).

24. Dreams of Earth and Sky

J. P. Ostriker and P. J. E. Peebles, "A Numerical Study of the Stability of Flattened Galaxies: Or, Can Cold Galaxies Survive?" *Astrophys. Journal, 186,* 467–480 (1973), were the first to state clearly the paradox of galactic stability and the hypothesis of invisible stabilizing matter.

For Elijah, see I Kings 19: 4, 11, 12.

INDEX

About the Author

Freeman Dyson has been a professor of physics at the Institute for Advanced Study in Princeton since 1953. Born in England, he came over to Cornell University in 1947 as a Commonwealth Fellow and settled permanently in the U.S. in 1951.

Professor Dyson is not only a theoretical physicist; his career has spanned a large variety of practical concerns. His is a unique career inspired by direct involvement with the most pressing concerns of human life, from minimizing loss of life in war, to disarmament, to thought experiments on the expansion of our frontiers into the galaxies. Professor Dyson has been a consultant to various parts of government—in particular the weapons laboratories, NASA, the Arms Control and Disarmament Agency, and the Defense Department. He has written articles for *Scientific American* and *The New Yorker* and has been awarded a number of distinguished prizes, among them the Hughes Medal of the Royal Society, London, the Max Planck Medal of the German Physical Society, the Robert Oppenheimer Memorial Prize, and the Harvey Prize by Technion, Haifa, Israel.

In addition to his scientific achievements, Professor Dyson has found time for raising five daughters, a son, and a stepdaughter.